The Nuclear Environmentalist

Juan José Gómez Cadenas

The Nuclear Environmentalist

Is There a Green Road to Nuclear Energy?

Copernicus Books

An Imprint of Springer Science+Business Media

Juan José Gómez Cadenas
Consejo Superior de Investigaciones
Científicas and Universidad de Valencia
Valencia
Spain

ISBN 978-88-470-2477-9   ISBN 978-88-470-2478-6   (eBook)
DOI 10.1007/978-88-470-2478-6
Springer Milan Heidelberg New York Dordrecht London

Library of Congress Control Number: 2011940858

Translator: Anahí Seri

Copernicus Books is a brand of Springer

Springer is part of Springer Science+Business Media (www.springer.com)

It's a black and white photograph, but it captures the light of a summer afternoon in 1959.

The man has a plucky nose, an honest chin, and a moustache in the style of Clark Gable.

The girl is a very beautiful brunette. Her smile is ecstatic; his, incredulous. Both of them, in love.

This year we're celebrating their golden wedding.

To my parents

Acknowledgments

Every book is a voyage.

Without my wife Pilar and my children Irene and Hector, it would not have been a voyage but a wreckage.

Without the help of numerous friends and colleagues, it would have been much more difficult to find a safe harbor. The list is long and the memory of the old sailor weak. Thus, I prefer to extend here my acknowledgment to all of them without spelling their names. You all know.

I am grateful for the kindness of the Spanish Nuclear Council and the Foro Nuclear, who have supplied information concerning Spanish nuclear power stations.

Contents

Chapter 1
All that Glitters is not Green

Oxymoron (from Greek oxymoros, "pointedly foolish"). A combination of contradictory or incongruous words, as cruel kindness. *(Merriam-Webster).*
Oxymoron (from Greek "sharp dull"). A figure of speech that combines contradictory terms. (Wikipedia).

The painter, carrying his easel, walks leisurely across the meadow that extends up to the limits of the summer sky. Under a chestnut tree he prepares his palette and his colors, then stretches and smiles. He is wearing a cotton shirt and slacks; a straw hat covers his curly hair. He walks barefoot because he likes the feel of the grass under his feet. This painter loves nature; he loves nature as an artist and as a scientist. This painter is a nuclear physicist, and his job consists in harnessing the elementary power of the atom, the one that makes the stars glow, in order to generate the power and hydrogen his town uses.

The town where the painter lives extends on both sides of a wide river, a few kilometers from this meadow. Today is the summer's solstice of the year 2050, and also the tenth anniversary of the Day of Change, the historic date when the last coal-fired power station was closed down. To celebrate this, a lot of families have gone on cycling tours along the car-free road lined by wind turbines that leads to the great reservoir, all of which, together with the nuclear power station, provide electricity for the homes and industries in town. Others, like the painter, practice their hobbies.

The town and its people reject excesses, detest wasting and believe in solidarity. They know it is necessary in order to improve a world which now, halfway into the century, already houses nine billion people. The people who live here consume less energy than they used to waste at the beginning of the 21st century: they live in highly efficient buildings, travel on high-speed trains, drive little hybrid cars. The sheer mention of the monstrous SUVs that used to cram the highways some decades ago sends shivers down their spines. However, more energy is consumed on the planet than ever before, as for the first time, all of its inhabitants have the right to a reasonable minimum.

Generating all this energy without the resort to the fossil fuels whose threat still lingers over the future like a Nazgul's shadow—the CO_2 concentration has stabilized at 450 ppm, and scientists hold the hope that a catastrophe has been averted—requires a momentous effort. The painter is proud of his work because he knows that it is an important part of this effort. Without him and many others like

J. J. Gómez Cadenas, *The Nuclear Environmentalist*,
DOI: 10.1007/978-88-470-2478-6_1, © Juan José Gómez Cadenas 2012

him, this meadow where he is strolling might be a bleak wasteland stricken by draught.

Today, the painter feels inspired. He fixes his gaze on the twin towers that dominate the horizon and gets to work. A while later, the giant chimneys of the nuclear power plant start to become visible on his canvas, but he has transformed them into huge trees, covered with green leaves.

Gaia

Gaia. During my first years at university, that was all we talked about. Gaia was Mother Earth, the living planet, the Earth Goddess made divine through science. And James Lovelock was her prophet.

Lovelock was working for NASA in 1965, engaged in a project that tried to find life on the Red Planet, when he realized that the atmosphere of Mars and Venus, like the one of the primitive Earth, was almost completely made up of CO_2. What had happened on our planet that had turned its atmosphere into something so different from its neighbors? The audacious scientist dared to postulate the hypothesis that life itself was responsible for these deep changes.

Lovelock liked to talk about Gaia as if it were an intelligent being, capable of globally controlling its own temperature, atmosphere composition and ocean salinity through, and in benefit of, living organisms. It is a beautiful and not quite accurate metaphor that has been very controversial in scientific circles, where poetic license is frowned upon, but which has also won him myriads of supporters. For all my generation, James Lovelock was not just an ecologist, but the incarnation of ecologism.

Few could compare to him in the shrine of our admiration. One of these was Carl Sagan, author of wonderful books dealing with the solar system, supernovae, the search for extraterrestrial intelligence, quasars, black holes and all the other prodigies the sky is teeming with. And then Isaac Asimov's novels were our gospels. Lovelock inflamed our spirit with the idea of a living planet. Sagan bewitched us with the beauty of the cosmos. But Asimov persuaded us that one day our ships would navigate this infinite sea, the universe.

Asimov's spaceships, needless to say, were powered by nuclear energy. There was no other way to reach the high acceleration which is necessary in order to travel at near-light speed. There was no other way to generate the electricity, the hydrogen, the food and synthetic materials needed by those oversized spacecraft which mankind boarded en route to the stars. There was no other way to feed the formidable magnetic shields protecting the fleet from high-energy cosmic rays. Like on Captain Nemo's Nautilus, those space vessels were driven by just one reliable, powerful agent. The atom.

The Threat of Climate Change

Three decades have passed since then. Asimov and Sagan are no longer with us, but 90-year old James Lovelock is as energetic as ever and still fond of metaphors, as shown in the title of his latest work.

In *The Revenge of Gaia* (Lovelock 2007) the old ecologist argues that humankind's lack of respect for the planet—which can be seen in the destruction of rainforests and biodiversity, together with the inordinate consumption of fossil fuels—is driving the Earth's capacity to counter the effects of greenhouse gases to the limit. The result can be frightening:

> "The planet we live on has merely to shrug to take some fraction of a million people to their deaths (referring to the December 2004 tsunami). But that is nothing compared with what may soon happen; we are now so abusing the Earth that it may rise and move back to the hot state it was in 55 million years ago, and if it does, most of us, and our descendants, will die."

Venus, whose size and distance from the sun are not very different from the Earth's, is a near example of how this announced revenge can strike. The enormous build-up of CO_2 in its atmosphere causes an extremely strong greenhouse effect, and surface temperatures rise to nearly 460°C. Venus is an inferno drowned in darkness. Light cannot pierce the thick layer of toxic clouds, composed of sulphur dioxide and sulphuric acid.

What forces hold the greenhouse effect at bay and spare us the fate of our ruined stellar twin? Lovelock maintains it is above all the biomass, forests, plankton and algae we are hurrying to destroy while we increase the CO_2 concentration in a suicidal way by burning coal, oil and natural gas. In his view, consequences will be devastating.

The IPCC's Forecasts

Lovelock is not the only scientist to hold this opinion . The recent report by the Intergovernmental Panel on Climate Change (IPCC 2008)[1] uses a more moderate and quantitative language, but reaches essentially the same conclusions, to wit:

- The concentration of greenhouse gases has increased exponentially since the beginning of the industrial age, particularly along the 20th century (Fig. 1.1).
- The release of greenhouse gases into the atmosphere has caused the average global temperature to rise by around one degree in the last one hundred years.

---

[1] The Intergovernmental Panel on Climate Change (IPCC) is a scientific intergovernmental body created by the World Meteorological Organization (WMO) and the United Nations Environment Program (UNEP). It is made up of hundreds of scientists from all the world, with the goal of studying climate change and its consequences.

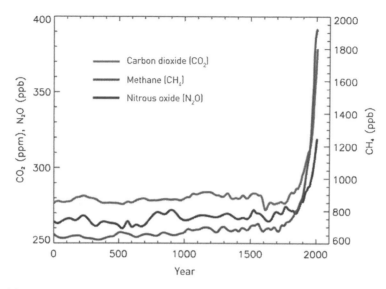

Fig. 1.1 Atmospheric concentrations of several greenhouse gases in the last 2,000 years (parts per million on the *left*, parts per billion on the *right*). The exponential increase of CO_2 is due to the human effect since the beginning of the industrial age. *Source* (IPCC 2008)

More specifically, since the middle of the last century, it has risen by half a degree, coincidentally with the increased concentration of greenhouse gases.

- At the end of the present century, the Earth may have increased by between one and three degrees. In the last case, consequences can be dramatic for our civilization: rise of sea levels flooding coastal cities, expansion of subtropical deserts, etc.

When Crocodiles Swam in the Arctic

Ours is not the first warm period in the history of Gaia. There was a similar one about 55 million years ago, at the beginning of the geological epoch known as Eocene, brought about by the release, in a brief lapse of time (between a few decades and two or three centuries), of billions of tons of CO_2 into the atmosphere.

What natural phenomenon could have given rise to such an increase in gases which under normal conditions are kept at constant concentrations on Earth? A possible explanation, due to the Norwegian physicist Henrik Svensen and his team (Svensen et al. 2004), points to the dissociation of methane hydrates, triggered by underwater eruptions in the North Atlantic, a region active at the time.

Methane hydrates are formed by compounding water and methane under high pressure and relatively low temperature, conditions which are prevalent in the deep sea. There are huge amounts of this compound, formed from decay of plankton and other organic matter. In a sense, this is one of the many mechanisms by which Gaia regulates herself; we can describe them as gigantic carbon deposits sequestered by living creatures in the sea.

We should not forget that the dreaded greenhouse gases are essential for life. One third of the solar radiation that hits the Earth is reflected back into space (by clouds, snow layers, oceans, etc.), the rest is absorbed and emitted back as infrared radiation and then again partially absorbed by gases which are present in the atmosphere in low concentrations, among them CO_2, methane (CH_4) and water vapor. However, the gases which make up most of the atmosphere, oxygen and nitrogen, do not absorb infrared radiation. Were it not for the CO_2, methane and water vapor, among others, the mean temperature on Earth would be about $-20°C$ instead of the cozy $15°C$ we enjoy on the surface of our planet.

In the words of Lovelock, Gaia "knows" how to keep the concentrations of greenhouse gases in the optimal range for life. At an average temperature of fifteen degrees, the sea is a good habitat for algae and other sea organisms that synthesize chlorophyll, sequestering any excess CO_2 from the atmosphere and taking it down to the sea floor when they die. If the concentration of CO_2 increases, so does the capacity of the algae to synthesize chlorophyll, making them thrive, and this in turn regulates the CO_2 levels by storage in the sea, for example in the form of methane hydrates.

But even the Earth can suffer from fever occasionally. At the beginning of the Eocene, this fever was caused by an escalating volcanic activity which made the ocean temperature soar and reversed the CO_2 capture cycle through methane hydrates. When these compounds decay, huge amounts of carbon are released to the atmosphere, which in turn increases the ocean temperature and breaks down more and more hydrates. This is akin to a disease, but our planet is very tough and soon after it found a new stable state (or rather a "metastable" one, in the sense that it is one among many possible states). During this new state, which in fact lasted for only a blink of the eye in geological terms, just one or two hundred thousand years, the temperature of the Arctic Ocean was 23 degrees, turning it into a comfortable habitat for species such as the crocodiles.

Being a good mother, Gaia loves all her children equally. Geological studies suggest that in those times there were tropical rainforests reaching up to a latitude that today corresponds to the north of France or the state of Maine in the USA. A lot of species would undoubtedly thrive in such a warm climate. Others would go extinct. In the words of Lovelock:

> By 2040, parts of the Sahara desert will have moved into middle Europe. We are talking about Paris. As far north as Berlin [...]. If you take the IPCC predictions, then by 2040 every summer in Europe will be torrid. It is not the death of people that is the main problem; it is the fact that the plants can't grow. There will be almost no food grown in Europe [...]. We are about to take an evolutionary step and my hope is that the species will emerge stronger. It would be hubris to think humans are God's chosen race.

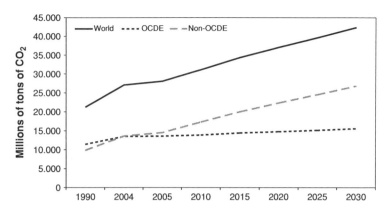

Fig. 1.2 CO_2 emissions into the atmosphere (historical and foreseen by EIA). *Source* (EIA 2008)

CO_2 and Fossil Fuels

In contrast to what happened in the Eocene, the current increase in the levels of greenhouse gases is not due to natural causes but to a strange bipedal, hairless, big-headed species which appeared recently on our planet and has even more recently started to burn fossil fuels in such huge quantities that the effect is comparable to the volcanoes of the Eocene. Figure 1.2 shows the global CO_2 emissions in millions of metric tons, for the world as a whole and for OECD[2] and non-OECD countries. It is striking to see that developing countries catch up with developed countries around 2005 and by 2030 emit 2.5 times more CO_2 to the atmosphere than the latter.

In 1990, petroleum was the main producer of CO_2 emissions (42%), followed by coal (39%) and natural gas (19%). In 2030, according to the forecast, coal is first (44%), with petroleum second (35%) and then natural gas (21%). The spectacular increase of emissions linked to coal (and to a lesser degree to natural gas) is basically due to the increase in electrical power generation.

Playing with Fire

According to the study by Svensen and his team, the volcanic eruptions in the Eocene may have released about 6 gigatons[3] to the atmosphere during a period of time ranging between 35 and 350 years. This amount is similar to what has been released since 1990 as a direct consequence of human action.

---

[2] The Organisation for Economic Cooperation and Development (OECD) is made up of 30 countries, mostly developed countries; its aim is to stimulate economic progress and world trade.

[3] A gigaton or Gt is a billion tons, see Chap. 2.

The reasoning is straightforward. If the volcanoes of the Eocene brought about the destabilization of carbohydrates, can't this happen again today?

Not immediately. The oceans are not yet warm enough for this to happen. However, even if we were able to stop CO_2 emissions dead in their tracks right now, the planet would keep on heating up for centuries. The current CO_2 levels (380 parts per million, ppm) are already above any maximum level in the past interglacial periods and their effect can be likened to a time bomb that will set off a retarded explosion. The oceans have already started to warm up, and if this goes on long enough, in a few decades or one or two centuries we will be going through the Gaia methane hydrates experiment revisited. The bomb has been activated, and a species wiser than us would be doing everything possible to defuse it before is goes off.

Nuclear Ecologists

In his work *The Revenge of Gaia* Lovelock does complain, but he does not leave it at that. He suggests urgent measures to stop CO_2 emissions before climate change turns irreversible. And he, the father of modern ecologism, makes a case, above all, for nuclear energy.

I am a Green and I entreat my friends in the movement to drop their wrong-headed objection to nuclear energy.

Lovelock is not the only ecologist to hold this view. Patrick Moore, one of the founders of Greenpeace and president of the NGO in Canada for years—though he would later leave this organization to found another group, called Greenspirit—shares his opinion. Even more noteworthy, the association Environmentalists for Nuclear Energy,[4] headed by engineer Bruno Comby, is a do-or-die advocate for the apparently blasphemous idea that nuclear energy is necessary for a better world. In the ranks of scientists, the supporters of nuclear energy are in the majority.

In contrast, organizations like Greenpeace are staunch opposers of everything atomic and have recently launched a harsh anti nuclear campaign in Spain, which is riddled with news that are unsourced, exaggerated or just plain false.

Who is right? In order to reach an unbiased opinion, you need some detailed knowledge of this fascinating subject. I challenge the reader to answer the following questions—without resorting to Google—and only then to check the footnotes.

---

[4] Which has inspired the title and the cover of this book, see http://www.ecolo.org/.

1. What releases more radioactivity into the atmosphere, a nuclear power station or a fuel or coal-fired power plant?[5]
2. What entails a greater risk, living next to a nuclear power station or smoking a cigarette?[6]
3. Isn't it true that the consumption of coal is going down across the world?[7]
4. Isn't saving enough to deal with the problem?[8]
5. Do nuclear power stations release CO_2?[9]
6. How much uranium do you need to generate as much energy as a ton of coal?[10]
7. How many wind turbines do you need to replace a nuclear power station?[11]
8. What happens during times of peak electricity demand if the wind doesn't blow?[12]
9. How do the building costs of photovoltaic solar parks compare with building a nuclear power station?[13]
10. A nuclear power station generates highly radioactive waste. What is the amount, in volume, of the waste produced by a typical 4 people family in Europe during all of their lives?[14]
11. How deep must they be buried so they don't have any harmful effects?[15]
12. Isn't it true that radioactive waste remains active for millions of years?[16]
13. Isn't it true that there is little uranium left?[17]

---

[5] A fuel or coal-fired power plant Chap. 9.

[6] Smoking *just one cigarette* entails the same risk as living next to a nuclear plant for two years Chap. 10.

[7] Quite on the contrary, it is growing dramatically Chap. 4.

[8] Not at all. Coal consumption and CO_2 emissions are especially high in developing countries, such as China and India, whose per capita consumption is much lower than ours, offset by a population of almost 3 billion people Chaps. 4 and 7.

[9] Direct emissions are zero. "Indirect emissions", related to their construction or to uranium mining, are lower than for photovoltaic or thermo solar plants, and in any case ridiculously small Chap. 11.

[10] Ten grams, in bulk equivalent to a pinhead Chaps. 9 and 11.

[11] Around two thousand latest generation models. If you place them 500 m apart, as needed to be efficient, the row of wind turbines would stretch from Barcelona to Geneva crossing all of France Chap. 12.

[12] You have to resort to hydropower or to "reserve" gas plants. Electrical energy can't be stored.

[13] As of today, a photovoltaic park is 10–20 times more expensive, per kWh, than a nuclear plant Chap. 12.

[14] A golf ball Chap. 9.

[15] A few meters depth is enough Chap. 9.

[16] A small percentage of the substances that accumulated in spent fuel have long half-lives. However, after a few thousand years, the activity of the waste is lower than natural uranium radioactivity in a coalmine. Besides, the waste with longer half-lives can be recycled and burnt with fast neutrons reactors Chap. 9.

[17] It depends on what you consider little. There's enough for about seven million years if we use it lavishly Chap. 11.

14. What's the need for nuclear energy? We can get all we need from renewable sources.[18]

If you have come close to the correct answers, you are either extraordinarily smart, or you belong to a minority of people who have a reasonable understanding of the pros and cons of nuclear energy. These questions, and many more, will be dealt with in what follows.

In brief: today, few people doubt that the most important concern of our times is how to avoid a global catastrophe due to climate change. However, serious as it is, the threat is not immediate enough for people, policy makers and those who still call themselves ecologists to be seriously alarmed. There's the paradox that we are still worried about the likely radioactivity of nuclear waste ten thousand years from now while we should be much more concerned about the explosion of methane hydrates in a century. In our days, being an environmentalist can't be synonymous with repeating the worn slogans over and over again and sticking to fanatic dogmas. All that glitters is not green.

How to Read this Book

This book is about energy, so it is worthwhile to start by reviewing the meaning of this term, which we all understand but few of us are able to define precisely, and by explaining the units used to measure it (Chap. 2). Five chapters devoted to understanding our society from the point of view of energy follow. We are absolutely dependent on fossil fuels (Chap. 3), and there is no way to understand the dilemma we are in unless we have a grasp of the history and the current situation regarding coal (Chap. 4), oil (Chap. 5) and natural gas (Chap. 6), all of which, but especially coal and gas, are used to generate the vital fluid that runs through the veins of our times: electricity (Chap. 7).

The second part deals with nuclear energy, one of the few alternatives left to us to avoid the disaster predicted both by Lovelock and the IPCC. Its history, one of the most enthralling of the 20th century, is told in Chap. 8. I also talk about nuclear reactors, explaining how they work and the reasons why they are safe (Chap. 9). I take a look at how the fear of radioactivity, accidents and terrorist attacks are justified (Chap. 10). And of course I address the touchy topic of waste. Finally, I shed some light on matters such as the abundance of uranium or the cost of nuclear energy (Chap. 11).

One of the points that is often made is that nuclear energy is unnecessary because we have renewables. This hypothesis is examined in Chap. 12. The last chapter glimpses into the future, wondering if there is a way to solve this mess we have been creating for a century.

---

[18] That's wishful thinking. The solar dream is still impossible, for reasons both physical—variability of sunshine—and technological and economical.

The future. Our grandchildren, or perhaps our grandchildren's grandchildren, will not resign to keeping chained to Gaia. Children grow up and leave home, and so will ours, one hundred or one thousand years from now, heading first to Mars and then who knows where. They will be few at first and a great crowd as time goes by. To travel, to know, to discover, it's in our nature. When they leave, they will do so in spaceships that have nothing in common with those imagined by the science fiction authors of my teenage years, except for one small detail: they will be powered by the atom.

References

EIA Energy Information Administration (2008). Annual energy outlook. http://www.eia.doe.gov/oiaf/aeo/.
IPCC Intergovernmental Panel on Climate Change (2008). http://www.ipcc.ch/about/index.htm.
Lovelock, J. (2007). *The revenge of Gaia: Earth's climate crisis and the fate of humanity*. UK: Penguin Books. ISBN: 0141025972.
Svensen, H. et al. (2004). Eocene global warming. *Nature, 429*, 542–545.

Chapter 2
Eternal Delight

Sisyphus in Hell

On his trip to Hades, Odysseus (Odyssey, XI) meets Sisyphus, King of Ephyra, who might be his illegitimate father. Like Odysseus himself, Sisyphus is a great sailor and an even greater liar; the father and his unrecognized son are both extremely cunning. Sisyphus's greatest feat is capturing Thanatos, the messenger of death, when he comes for him, thus upsetting the world, as nobody dies during some time until Ares manages to fix the mess. When the great deceiver finally ends up in Hell, he is compelled to roll a huge bolder up a hill. As soon as he reaches the top, the boulder slips from his sweating hands and rolls back down to the valley. Sisyphus is forced to repeat the same drill throughout eternity. (Fig. 2.1)

In order to roll up the boulder, Sisyphus has to apply (muscular) energy to counter the force of gravity that opposes his efforts. As a result of his work, when the rock has reached the peak of the hill it has gained a kind of energy we call *potential energy*, E_p. The rock is able, it has the *potency* (hence the term "potential") to carry out some work while rolling down, and this work is proportional to the mass of the rock (m), the height of the mountain (h) and a fixed value that stands for the action of gravity (g), that is, $E_p = m \times h \times g$.

Sisyphus transforms his muscular energy into potential energy, which can in turn be transformed into electricity: if he had been condemned to push up a large water container instead of a rock, the water running down could have powered a turbine connected to an alternator to generate electricity. In the whole process there is a flowing quantity whose magnitude remains unchanged while its quality is transformed (muscular, potential, electrical energy). *Energy can neither be created nor destroyed: it can only be transformed.* This is the first and most famous law of thermodynamics, formulated by the great English physicist J.P. Joule (1818–1889) after years of time-consuming experiments, based on the observations by the German physician and physicist J.R. von Mayer (1814–1878).

J. J. Gómez Cadenas, *The Nuclear Environmentalist,*
DOI: 10.1007/978-88-470-2478-6_2, © Juan José Gómez Cadenas 2012

Fig. 2.1 Sisyphus rolling the boulder uphill. As it rolls down, it is able to perform work. We express this by saying that the boulder gains *potential energy*

Power

The concept of power is as familiar to us as the concept of energy, but we often mistake one for the other. The correct definition of power is *the capacity to do work per unit of time.*

Let's take the example of two Sisyphuses toiling up the mountain, each with his rock, both of equal weight. One of them, more able-bodied than the other, manages to push the rock up at a faster pace (that is, he performs more work per unit of time, in other words, he develops more power), so he overtakes his fellow sufferer. Both, as we know, receive an identical reward: when reaching the top, the rocks slip from their hands. Both rocks are capable of doing the same work, so both convicts have generated (and consumed) the same amount of energy. The brawnier Sisyphus has a greater power, but this just means that he is able to do the work faster than his feeble fellow.

It is important to realize that in order to relate the power generated or consumed by a process to the amount of energy consumed we have to resort to time. A stupid little example: which car consumes more energy, a small 100 Hp car or an SUV with 500 Hp? The obvious answer: *it depends on how long the engine is running.* All of the power of a Mercedes Benz does not use up a single drop of oil unless we start a car (but of course is doesn't take us anywhere).

Units for Measuring Energy and Power

Energy is measured in different units, of which the most common in everyday life is the kilocalorie, which stands for the amount of energy you get from food. Everybody knows, for example, that the amount of energy an adult person needs daily is between two thousand and three thousand *calories,* depending on sex, age, build and activity level (a moderate diet for weight loss would allow about 1500 calories per day, and there are rapid weight loss diets where you have to limit yourself to 1000 calories).

Sounds familiar, doesn't it? It's wrong, too. A 3000 calorie diet wouldn't keep a 20 g mouse alive. When we use the word "calorie" we mean "kilocalorie", that is, one thousand calories. Thus, the average amount of energy we need is around 2500 kilocalories, that is, 2500×1000 calories, this is 2,5 *million* calories, in short 2,5 Mcal.

The calorie is a common unit but does not belong to the so called International System of Units or SI, which includes the meter as unit of length, the kilogram as unit of mass and the second as unit of time. In the SI, the unit of energy is called Joule (in honor of the physicist J.P. Joule) and is represented by the symbol J. A calorie amounts to 4.18 J, so our 2500 kilocalorie (2.5 Mcal) diet represents an energy of 10.5 million Joules, or 10.5 MJ.

The Joule, the same as the calorie, is used to measure small quantities of energy, that's why we employ prefixes to make the numbers more manageable. Instead of speaking of an average 2,500,000 calorie diet, we say 2,500 kilocalories or 2.5 *Mega* calories. The same happens with the Joule. The most common prefixes are given in the following table.

Prefix	Symbol	Value	Decimal	Example (Joule)
Kilo	k	One thousand	10^3 (1,000)	kJ
Mega	M	One million	10^6 (1,000,000)	MJ
Giga	G	One billion	10^9	GJ
Tera	T	One trillion	10^{12}	TJ
Peta	P	One quadrillion	10^{15}	PJ
Exa	E	One quintillion	10^{18}	EJ

Some examples: a pea contains 5,000 J (5 kJ) of chemical energy. A mouse needs about 50,000 J (50 kJ) a day, an adult man approximately 10.4 kJ. The oil tank of a passenger car holds around 1.25 GJ.

Figure 2.2 shows the energy yield for different fuels. We can see that one kilogram of hydrogen is equivalent to two and a half kilogram of petrol, three of natural gas, seven of wood and ten of straw or dung. Considering fossil fuels, oil is the most energetic: one kilogram provides as much energy as two kilogram of coke, three of wood or four of straw.

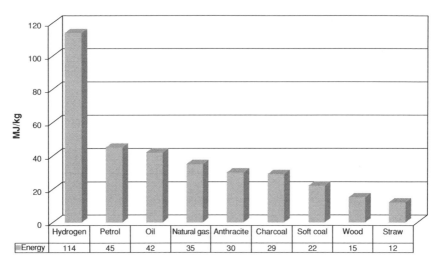

Fig. 2.2 Calorific power of different fuels

A unit of energy that is used quite often is the ton of oil equivalent or *toe*. Its value is the amount of energy released by burning one ton of oil. If one kg provides 42 MJ (Fig. 2.2), from one ton you get thousand times as much, that is, 42 GJ. This unit allows us to compare several fossil fuels in terms of energy. For example, 1 ton of natural gas is equivalent to 0.83 toe, 1 ton of anthracite is equivalent to 0.7 toe and one ton of coke is equivalent to 0.52 toe.

Unlike the (kilo)calorie, the most well known unit of power, the watt (W), does belong to the SI. Its name honors James Watt (the inventor of the first modern steam engine) and is defined as the work of one Joule per second (that is: 1 W = 1 J/s). When we say that a light bulb has a power of 100 W, we mean that in order to keep it lighting we need 100 J of electrical energy per second. So, if the bulb remains on for 5 h a day, the energy it consumes per day is 5 × 60 × 60 × 100 = 1,800,00 J or 1.8 MJ. Curiously enough, the basal metabolic rate of a stout adult male is about the same, around 100 W. To find out how much energy this metabolism consumes in a day we have to multiply by 24 h because the basic chemical processes that keep us alive are switched on all the time. So that's 24 × 60 × 60 × 100 = 8,640,000 or 8,6 MJ.

We mustn't confuse the kilowatt (kW), a unit of power (work per unit of time), with the kilowatt hour (what we are charged for in the electricity bill). The kilowatt hour (kWh) is a unit of energy which results from multiplying the power of one kilowatt by the time of one hour and is equivalent to 3.6 MJ. We can see that it measures larger quantities of energy than the Joule and can be more convenient. The typical energy consumed by a European family that uses electricity for lighting and household appliances (but nor for heating and air conditioning) is around 250 kWh a month (about twice as much in the US). By adding heating, air conditioning and an electrical stove, this goes up to about 500–1,000 kWh a month.

Finally, there is another common power unit not included in the SI: the horsepower (HP), which we still use to refer to the power of automobiles and which is literally a measure of the power of a draft horse. People used to compare the first steam engines with these horses. When we say that our car has 100 HP we are literally referring to a herd of one hundred horses pulling our vehicle, and to their capacity to perform the work per unit of time, though one century ago, few people would have been wealthy enough to afford the stables and the grain needed to feed such a bunch of animals. A horsepower of one is equivalent to 745 W.

Entropy and Dark Energy

The so-called second law of thermodynamics was formulated by the German physicist Rudolph Clausius (1822–1888), who in an article published in 1865 coined the term *entropy*, defined as the disorder of an isolated system. The second law of thermodynamics can be expressed in a very condensed but a little cryptical form:

The entropy of an isolated system increases continuously.

In plain language, this means: *In an isolated system the amount of available energy to perform work becomes smaller and smaller over time.*

A straightforward example: before burning, a piece of coal holds "high quality" energy due to its very organized crystal structure. So its entropy is low. Once the coal has been burnt, the energy it contains does not disappear, but is transformed into heat, a very disorganized (high entropy) form of energy. The total energy of the system remains the same, but once the internal energy of the coal has turned into heat it cannot be used again to produce useful work. That's the reason why a *perpetuum mobile*, or perpetual motion machine, will never work, however ingenious the design may seem. Every engine produces heat because of the friction of the parts and therefore energy is continuously dissipated, which leads to a standstill of the engine if there is no provision of fuel. In fact, heat occupies a peculiar place in the scale of energies. Any kind of energy can be turned into heat, but heat itself cannot be converted into any other kind of energy.

On the other hand, our common experience tries to persuade us that the second law of thermodynamics does not hold. To begin with, living creatures seem to violate it at all stages, from the moment of conception and the development of individuals (where a disorganized bundle of cells organizes into something as extremely orderly as a human being), to the evolution of species, which seems to progress from the simple (unicellular animals and plants) to the complex (men and angels). And then, how come there are renewable energy sources, if the increase of entropy should do away with them? How is it possible that the wind keeps blowing? Shouldn't the second law of thermodynamics deprive us of this useful energy? The answer to both questions is the same. Our planet is not an isolated system, but an open one, receiving a continuous flow of energy from the sun. This is the energy that plants profit from in order to create biomass through photosynthesis; the energy that generates the winds that move the wind turbine blades, the energy nature makes use of to move the unrelenting machine of evolution.

However, the universe is by definition an isolated system, so the second law of thermodynamics predicts its famously tragic thermal death. As time passes, the immense energy released by the Big Bang is being transformed into nebulae, galaxies, stars and living beings. Unfortunately, it doesn't end there. Eventually the stars will go out, galaxies will move apart from each other, and the universe will be thrown into disarray. And as the universe expands the particles it is made up of become cooler and cooler, until the moment of maximum disorder arrives, and with it the cold, the most absolute solitude.

Until recently we physicists believed there was another possible *Grand Finale*, with the universe contracting again, pulled by gravity, inverting the second law of thermodynamics, turning on the stars, forming ever tighter and denser cumuli finally leading to the initial singularity that created us. The latest observations seem to suggest otherwise. There is something, a force we don't understand and which rushes to push the universe into continuous expansion and thermal death. For want of another name, we call it Dark Energy, an expression that in fact might be appropriate, given the end it hurls us against. It has appeared rather recently (given the time scale of the Universe) and to understand its origin is possibly the greatest mystery physics faces in the 21st century.

But that's another story.

Chapter 3
A Wasted Inheritance

A man had two sons. The younger son asked his father to give him his share of the estate. The father divided the property between the two sons. A few days after, the younger son took his things and traveled to a country far away. There he wasted all of his wealth, living foolishly.

Parable of the prodigal son, New Testament, Bible

Aladdin and the Genie

We all know how Aladdin escapes when the sorcerer traps him in the magic cave. He finds a lamp, rubs it, and a powerful genie appears, who will grant him all his wishes. To start with, he takes him home swiftly, though perhaps not as fast as if he had boarded an Airbus or a high-speed train. Then he loads his table with delicious food, almost as plentiful and varied as can be found in our refrigerators. Finally, he dresses him in luxurious silk clothes, like the ones you can get at the sales in Macy's. The young boy grows confident and relishes in his pleasant life, taking it for granted that he deserves everything he is profiting from.

But then the wicked sorcerer comes back to reclaim his property, and things get rough.

In the story I am about to tell you there is a magic lamp as well; a lamp which, as in some variations of the story, grants three and only three wishes: they are sufficient. Ask for coal, oil and natural gas, and the rest will be added unto you. Our society is as different from the ones that came before us, in the last ten thousand years, as Aladdin's house differs from the other houses in his wretched neighborhood.

All traditional societies have obtained light and warmth by burning wood, bush, straw and dung, relying upon the muscles of men and draft animals for housework, agriculture and building. Figure 3.1 shows the contribution of various primary engines along history. Until about two hundred years ago, the main resource for carrying loads, hauling supplies and performing heavy tasks such as plowing, grinding or lifting weights were human and animal muscles. Not until the 13th century did the first mechanical devices (windmill and water wheels) start to play a significant role in Europe, though their applications were limited and they did not do much in the way of easing the rough living conditions. In the 18th century, human and animal work still made up more than 85% of the total effort. Not until that astounding epoch at the end of the 19th and the beginning of the 20th century does man, for the first time, cease to be basically a beast of burden. From 1950 on, the

J. J. Gómez Cadenas, *The Nuclear Environmentalist*,
DOI: 10.1007/978-88-470-2478-6_3, © Juan José Gómez Cadenas 2012

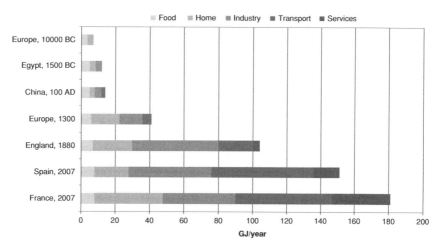

Fig. 3.1 Energy consumption in traditional and modern societies: adapted from (Smil 1994)

prominence of internal combustion engines (motorcars, tractors, ocean liners, oil tankers, trains), electrical engines (tram, subway, high-speed trains, Shinkansen[1]), mobile turbines (commercial airplanes) and gas turbines (power plants) have turned every citizen in the developed countries into a Croesus, an Ali Baba or a Rockefeller, definitely wealthier than any tycoon in ancient times.[2]

The amount of energy available to each person before the industrial revolution was small, and for centuries it increased very slowly. All through the Middle Ages famines were always around the corner; for heating and cooking, people depended on nothing more than a fireplace in a common room; in all their lives, few traveled more than 50 km away from their birthplace. Life expectancy was short, illiteracy was pervasive, leisure unheard of. The gentry were a bit better off, but not even the most powerful duke had access to X-ray screening that might detect a cancer in its early stages, not to speak of precious anesthesia to spare him the hideous pain of a simple tooth extraction.

Thanks to the availability of fossil fuels and the numerous technical advances associated to it, industrialized countries currently have huge quantities of energy at their disposal, as can be seen in Fig. 3.2, where France and Spain today are compared with several societies of the past: the hunter-gatherers from 10,000 years ago, ancient Egypt (still in the bronze age, but already with stable agriculture, irrigation systems and surplus energy to build the pyramids), the Han dynasty in China, 100 BC (an agricultural society with advanced irrigation projects and metal tools), medieval Europe around the year 1300 (able to forge steel and

---

[1] The famous Japanese Bullet Train.

[2] Here we might add "and as pricked by conscience as they were", considering that more than a billion today people live on less than a dollar a day.

Fig. 3.2 Total primary energy worldwide. *Source* (BP 2008)

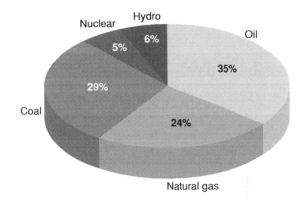

build gothic cathedrals) and finally England in 1880, an industrial society on the rise, fed by coal and driven by steam engines.

Coal made the industrial revolution possible, but oil was the fuel responsible for the revolution of transport and of primary motors, which—literally- keep the world moving. Let's picture an engineer in the Middle Ages, in charge of building a cathedral, a road or an irrigation channel. If he had at his disposal a work crew of a hundred strong men, the power available would equal a small tractor from the 1920s. A workforce of 2,500 individuals would amount to a modern tractor. The comparison is still more impressive if we turn to ships, considering human powered galleys versus present day diesel motor vessels: three hundred thousand galley slaves would be needed to drive one of these ships. Assuming the rowing power to be sufficient for flying, six hundred thousand oarsmen would have to be crammed into a galley to reach the power developed by the four turbines housed in a Boeing 747.

From oil we do not only obtain the petrol for our cars, the diesel oil that drives tractors, trucks and ocean liners, and the kerosene a Boeing needs, but quite literally everything around us. Plastic, paints, disinfectants, shoe soles, wheels, asphalt, glue, dyes, preservatives, electric tape, synthetic rubber, photo film, contact lenses, credit cards, insect repellent, washing powder, anti allergy drugs, toothpaste, perfume, lubricant, paint remover, PVC, lipstick, aspirin, anesthetics and computer chips.

The third fossil fuel is natural gas, which consists almost entirely of methane, a carbon atom bonded to four hydrogen atoms (CH_4), whose chemical structure is very simple compared with its relatives, compounds made up of long chains of carbon and hydrogen atoms. Natural gas, besides being the fossil fuel which releases the fewest CO_2, lacks other pollutants given off by oil and coal and can be used very efficiently for heating, electricity generation and even for transport. What is more important, it is essential for the Haber–Bosch process, which allows to synthesize the nitrates fertilizers are based on; without them, between a third and a half of the world's population would have starved to death in the 20th century.

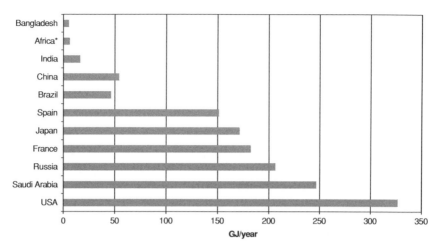

Fig. 3.3 Energy consumption in different countries in 2007. Africa* means the average of the African continent without Egypt, Algeria or South Africa. *Source* (BP 2008)

The Prodigal Son's Inheritance

These three treasures bear a resemblance, to express it in biblical terms, with the prodigal son's inheritance. The legacy nature bequeathed to us included next to one billion tons of coal, more than two hundred thousand million tons of oil and the same amount of natural gas, an immense energy stock that has provided the base and the sustenance of the greatest revolution in history, as can be seen in Fig. 3.3. 88% of the primary energy worldwide[3] is extracted form fossil fuels, leaving only a meager 6% for nuclear energy and for renewable energies respectively, with hydroelectric energy being the dominant among the latter.

Little more than two centuries have passed since James Watt's steam engine, a fleeting instant even at the time scale of human history. The last six generations have witnessed a dazzling succession of inventions and technical advances, many of which appeared in a lapse of not much more than one hundred years, and which have resulted in a complete transmutation of the world: from the steam locomotive (1814) to the high speed train; from the invention of Otto's explosion engine (around 1870) and the Diesel engine (1892) to the Ford-T (1920) and the Toyota Prius; from the Titanic to the modern oil tankers, the largest being more than ten times as heavy as the mythical transatlantic liner; from the Wright brothers' fragile biplane (1903) to the space shuttle.

And still, the transport revolution fades when compared to electrification, which was completed in developed countries before World War II and led to the spread of the electrical engine, electronics and finally computing. These wonders are

---

[3] Primary energy includes electricity, transport, home, industry and services.

accompanied by medicine based on the discovery of antibiotics (Fleming 1928), nuclear physics (X rays, radiotherapy, scans), molecular biology and genetic engineering.

In Europe, the USA, Japan and other rich countries a middle class has emerged that comprises most of the population, has access to all these technical miracles and is entitled to education and to health care. The average life expectancy has risen more than 50% in under a century, leading to an increase in literacy, labor rights, gender equality and quality of life. New York or Madrid would be enchanted places not just for a countryman from the Middle Ages but for a scientist from the early 18th century: places where almost everything—communication, transport, household tasks—happens by magic, where common people enjoy privileges and luxuries unimaginable for noblemen of other times.

This development has been possible thanks to the availability of huge amounts of cheap, easy-to-use energy stored in concentrated form in fossil fuels. But the use of these fuels is beginning to cause trouble. In 2007 we gobbled up six billion tons of coal, three billon tons of natural gas and four billion tons of oil. Like the prodigal son in our parable, we spend without restraint, and like in his case our days of debauchery are numbered. Cheap oil is running out or will run out sooner or later, and it is very likely that this will become noticeable in a few decades. With the shortage of oil, the economic crisis will haunt the global village, even more so if natural gas does likewise. And though coal may well last longer, it is the fuel that emits most CO_2, and thus the fuel that contributes most to the excessive greenhouse effect responsible for global climate change, with potentially catastrophic consequences.

Sustainable Development?

In 1800 there were about 900 million souls on our planet. In 2009 the world population is nearing 7,000 million and by the middle of the century the figure will be between 8,000 and 10,000 million. It has become fashionable to speak of "sustainable development", without realizing that this concept is something made up by rich societies. In fact, the 20% of the world population we belong to monopolizes 80% of its resources, both economical and energetic, an outrageous truth that can be appreciated in Fig. 3.3. While one fifth of mankind is devouring eleven million tons of oil a day—equivalent to the mass of two hundred ocean liners, or around twenty skyscrapers—a citizen of Africa or Bangladesh makes do with less energy than a hunter-gatherer ten thousand years ago, while people in India, China or Brazil aren't much better off than serfs in the Middle Ages.

Almost everybody in the rich countries agrees that the world we live in is unfair and immoral, but we often don't realize that in order to remedy this injustice we need, among other things, to get the majority of mankind out of the Middle Ages. Disregarding catastrophes, to imagine that the USA or the European Union will half their consumption or reduce it to a third is wishful thinking. It is true that in rich countries there is a tendency towards moderation in energy consumption, but this just implies a slower growth that might come to a halt or even decrease slightly. The rapid growth of emergent economies such as China more than makes up for this.

When we speak about sustainable development, or about protecting the environment, or the town of the future, with its intelligent buildings, renewable energy sources and electric cars, we forget the myriad of towns that far from being intelligent or even human, look like garbage dumps where dwellers crowd in shacks which are the antipode of the intelligence, energy efficiency and comfort we are used to. Similarly, when we imagine that saving energy will allow us to consume less and thereby lessen CO_2 emissions, we tend to overlook the fact that half of mankind is obliged to consume *more* in order to escape from poverty. I do not mean the obvious fact that Chinese and Indians desire cars and washing machines, leisure time and decent salaries like us. Energy is also necessary to ensure that the people of Bangladesh and all of Africa have access to electricity, water and sanitation.

There is a childhood memory burnt into my mind, my dad urging us children to finish off our plates. "It's not fair to throw away food while there's so many people starving," he would say, over and over again. I did not get the point. What would a famished Ethiopian child gain from our empty plates? But my father was right. Turning off the tap, substituting a hybrid car for the SUV and remembering to turn off the lights when leaving home not only saves energy and CO_2. It also helps us remember that there are a billion destitute people scraping a living on our planet.

References

Smil, V. (1994). *Energy in world history*. Boulder: Westview Press Inc.
BP (2008). BP World Statistics. http://www.bp.com/.

Chapter 4
The Ignoble Fuel

> *Just like bacteria, fungi, and higher animals, humans will*
> *always seek new sources of cheap, accessible organic carbon.*
> T. W. Patzek 1995

Perhaps the first historic mention of a substance derived from coal goes back to the Bible, when Yahweh orders Noah to build the famous ark:

> So make yourself an ark of cypress wood; make rooms in it and coat it with *pitch* inside and out.
>
> *Genesis*, 6:14.

The fact is that the use of charcoal can probably be traced back to the moment fire started to be used, to the scraps of charred wood left over at a campfire. There are proofs that in many of the cave paintings made more than 15,000 years ago charcoal was employed to outline shapes, and also used as black pigment mixed with fat, blood or fish glue. The Egyptians made pigments from charcoal and Hippocrates recommended it for medical use as early as 400 BC. The Romans used charcoal in their steelworks and were acquainted with mineral coal, probably after invading the British Isles, though it served mainly for ornamental purposes. The Chinese included it in the recipe for gunpowder; the British breweries, in their filters, to remove impurities from alcoholic drinks; Edison, in the filament of his first light bulb. In everyday life, we find products derived from coal all around us: in pencils, in the form of graphite; in battery electrodes; in the rechargeable batteries of our laptops; in tires, which derive their black color from it; in the carbon fiber composite materials that make up aircraft wings, aero generator blades and all kind of prostheses; in the activated coal filters that purify water and air. It is used to absorb odors, as a remedy for diarrhea, minor intoxications, flatulence and bad breath and most importantly in dialysis. It is necessary for the production of ink, shampoo, perfumes and high technology tools; the noble form of coal, diamond, is the chief of precious stones, and synthetic diamond, as hard as its natural cousin, but much cheaper, is used to make bore heads. The latest and most promising application is nanotechnology: carbon nanotubes and nanofoam are astonishing developments, literally the materials of the future.

J. J. Gómez Cadenas, *The Nuclear Environmentalist,*
DOI: 10.1007/978-88-470-2478-6_4, © Juan José Gómez Cadenas 2012

Charcoal

By charcoal we actually mean several different materials, all of them related to burning biomass under low oxygen conditions. While *mineral coal* is a sedimentary rock formed in a slow geological process called *carbonification* that takes hundreds of million years, *charcoal* has been obtained since ancient times (and is obtained today in Africa and other regions) by burning wood in primitive ovens.

Charcoal is an excellent fuel. As it consists of almost pure carbon, it has a high calorific value, providing twice as much energy as good quality wood. Furthermore it doesn't contain any pollutants such as sulfur or phosphorus (which are present in most mineral coals). The lack of pollution fumes makes it especially convenient for cooking and heating. Another traditional and very important use has been metal melting. Steel metallurgy, which began around 1,200 BC and started to develop in Europe in 700 BC wouldn't have been possible without charcoal, as the high temperatures needed to melt minerals cannot be reached using wood (in fact, the use of common coal is out of the question in modern blast furnaces because of its impurities). Besides, the carbon contained in charcoal reduces the metal oxides that make up the minerals, and with an appropriate technique, part of this carbon can be allied to the iron to produce steel. The use of charcoal in metallurgy has continued to our days. Other fuels, such as metallurgical coke, have replaced it almost completely in developed countries, but nowadays charcoal is being revived in developing countries that are rich in forests.

Biomass as Fuel

Contrary to what we might think, the use of biomass as fuel is not a recent idea that has emerged in industrialized societies interested in renewable energies. It's quite the opposite: renewable energy has been the only available energy for the most part of human history. From the first urban civilizations in Mesopotamia, around 3,200 BC, to the great cities of the 16th century, like London or Venice, the main energy source, both for domestic and industrial usage has been vegetable and organic matter, burnt either directly (wood, straw, sugar cane, waste, corn ears, roots, dry dung) or indirectly in the form of charcoal.

Until the Late Middle Ages, population density in Europe was low and forests were abundant, but from the 12th century onward supply started to decline. Biomass provides little energy *per surface unit*, about 300 MJ/m^2 for good quality wood, and only a tenth of this value if we use straw, dead leaves, shrubbery or charcoal, as traditional charcoal ovens have very low efficiency.

In medieval Europe, each inhabitant needed about 10 GJ per year for cooking and heating, so 30 m^2 of good forest was necessary per person if wood was used directly, and about 150 if it was transformed into charcoal. This means a city with a population of one million would have needed between 30 and 150 km^2 of forest *per year*, and would have ravaged between 300 and 1,500 km^2 in a decade. Not too

Fig. 4.1 The difference between Haiti and the Dominican Republic (Thanks to F. Camarena)

sustainable, to put it into contemporary speech. An obvious corollary is that using biomass as fuel ruled out the existence of large cities during the Middle Ages.

In our days, we can see an example of forest devastation in Haiti, whose economy based on the massive consumption of charcoal has destroyed the woods in that part of the island, in contrast with the neighboring Dominican Republic, where they are still mostly untouched (Fig. 4.1).

Charcoal (the same as wood, sugarcane, corn etc.) is a renewable resource. Until the excessive exploitation surpasses regeneration capacity in the area. This is a lesson we should not forget.

Mineral Coal

As opposed to charcoal, mineral coal is formed by *carbonification*, the process by which vegetable matter (leaves, wood, bark, and spores) is slowly transformed in the partial absence of air, in peat bogs or in the sea at shallow depths. The process is pictured in Fig. 4.2. Most deposits were formed during the Carboniferous period about 300 million years ago, a few in the Triassic and Jurassic, and in smaller quantities in the Cretaceous.

Coal in the Middle Ages

In the late Middle Ages, the population of the boroughs and cities which were beginning to form depended on firewood to cook and heat their homes. Wood was brought in from the surrounding forests, charcoal was also used, and from the 13th

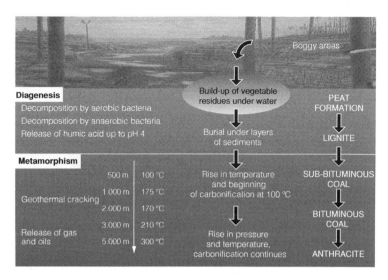

Fig. 4.2 Carbonification. *Source* (Menéndez 2008)

century on there was a supply of mineral coal from shallow mines and surface deposits which were very abundant in some regions, such as North East England. But while charcoal burns clean, "sea coal", as the mineral was called in England (it arrived in London in ships that used it as ballast), produced sulfur fumes that where intolerable in the primitive lodgings whose only exhaust was a hole in the roof. When making charcoal, more than half of the wood used in the process is wasted, so wood remained the fuel of choice, not only in households but also in the numerous factories that prospered at the time, among them distilleries and ship-yards. However, as the population grew, forests disappeared quickly, and by the middle of the 13th century wood had to be carried in from ever more far-away places. Scarcity increased the prices, and for the poor it became increasingly difficult to heat their homes.

Unfortunately, the problem of overpopulation in the Middle Ages was drastically solved in the 14th century, when the Black Death dispatched one-third of the European population. It is hard to imagine the scale of this catastrophe. In a few years, one in every three inhabitants of the old continent would die. Families, neighborhoods, whole villages were annihilated. There were corpses everywhere, too numerous to be buried decently, too common to inspire a feeling other than fear or weariness.

Then the plague passed, leaving a dwindled population and thousands of acres of deserted fields where forests thrived again, providing enough firewood for the surviving few.

A hundred and fifty years later, Europe went through the so-called "little ice age", a period that lasted from the end of the 16th century to the beginning of the nineteenth. The average temperature dropped up to one degree. In our days we fear global warming; in the 16th century people had to get to grips with an almost glacial climate: long, freezing winters that came along with a new population increase and new shortage of wood linked to deforestation. A miracle was called for in order to avoid a new catastrophe, and this miracle came with the extended use of mineral coal from the beginning of the 17th century onwards. The solution to the fuel scarcity came at a price—we can see that energy crises are not a modern invention either. In large cities such as London, the air was so polluted that on certain days the sun could hardly pierce through the dense fog it was covered in. A great relief came with the brick chimneys that started to spread, but the 19th century city, described by Dickens in the opening lines of "Bleak House", is a gloomy town, almost perpetually covered in smog, polluted and brutal.

> London. [...] Implacable November weather. As much mud in the streets as if the waters had but newly retired from the face of the earth [...] Smoke lowering down from chimney-pots, making a soft black drizzle, with flakes of soot in it as big as full-grown snow-flakes—gone into mourning, one might imagine, for the death of the sun. Dogs, undistinguishable in mire. Horses, scarcely better [...] Foot passengers, jostling one another's umbrellas in a general infection of ill-temper, and losing their foot-hold at street-corners, where tens of thousands of other foot passengers have been slipping and sliding since the day broke (if the day ever broke).

The Industrial Revolution

The great invention that finally made wood and charcoal lose its dominance to mineral coal was the steam engine, the heart of the industrial revolution, nourished by this fuel from the very beginning.

At the beginning of the 18th century, coalmines were threatened by frequent flooding that eventually rendered them useless. They were drained by chains of workers; several mechanic devices were introduced, such as windmills and waterwheels, and beasts of burden employed, but none of these techniques was convenient or economical enough.

At the beginning of the 18th century, an ironmonger called Thomas Newcomen invented a contraption that improved the situation. The device included a piston that was pushed up by the expanding steam generated by water heated with coal; then the steam was condensed with cold water and the piston came down again. The piston was connected to the axle of a pump used to drain the water. The machine was an immediate success, as it was much cheaper than the crews of workers or horses that had been employed up to then. However, Newcomen's engines were primitive and inefficient, and used up so much coal that they were unpractical for uses other than draining mines.

Things changed when James Watt, among other improvements, added a condenser to Newcomen's device and built the first modern steam engine (Fig. 4.3). Being more efficient, this engine could be taken out of the coalmines and installed in factories, where its impressive power allowed to multiply productivity while reducing the cost of human labor and animal workforce.

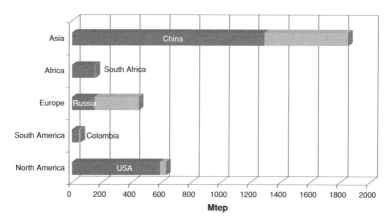

Fig. 4.3 Coal production per region, in tons of oil equivalent (toe). Next to each region, the main producer is given. *Source* (BP 2008)

But yet another technological advance was needed for industrial revolution to happen: the production of steel using metallurgical coke instead of charcoal. Metallurgy still depended on charcoal because mineral coal had too many impurities to forge metal with, and, as we have seen, there weren't enough woods to provide all the charcoal needed for mass production. The problem was difficult to solve and in fact it took almost a century of experimentation. The solution that was finally thought up was similar to the process by which charcoal is obtained: mineral coal is cooked in a way that removes volatile compounds and thus impurities, transforming it into coke, apt for the steel industry.

Great Britain, where all these inventions had taken place, embarked on a massive production of iron and built an industry that in a few decades had set the foundations of its mercantile and military power.

A third invention was added to the previous ones: the steam locomotive, created by George Stephensen, originally intended to transport coal to the new industry centers, Manchester and Liverpool. A few years later, England was covered in railways, and the train—as marvelous for people of the time as a spaceship is for us—began to take passengers and merchandise all around the country. When the rest of the European nations started to catch up with the industrial revolution, Great Britain was 50 years ahead, and this allowed her to create and consolidate her huge empire in the 19th century. However, in the 20th century she had to yield to another power with more natural resources and more work power, where the industrial revolution was still more explosive: the United States of America. The 21st century may well witness the emergence of new powers like China and India, which are already threatening America's supremacy in many areas.

Coal in the 21st Century

In Elizabethan times, London imported around 24,000 t of coal per year (Boyle 2003). In 1680, consumption had risen to 3.6 million tons, then 10 million around the year 1800 and 250 million in 1900. In 2007, a total of 6,395 million tons of coal (the energy equivalent of 3,135 million tons of oil) was mined all around the world, equivalent in weight to ten skyscrapers the size of the Empire State building. This enormous quantity amounted to one-third of the world primary energy (Fig. 3.2).

Figure 4.3 shows the world production per region, together with the greatest producer in each of them. 40% of the coal that was mined on the Earth in 2007 came from China and almost 20% from the USA. In Africa, 99% of the coal is mined in South Africa, and in South and Central America. 85% comes from Colombia. The largest producer in the European region (including the countries of the ex-USSR) is Russia, but on the old continent the resource is quite evenly distributed. Other countries holding substantial stocks are India (7.5%) and Australia (6%).

Coal is used above all for electricity generation in thermal power stations, which are just giant kettles heated with coal where very hot, high-pressure steam is produced, which then spins a turbine that drives an electrical generator. On average, 40% of the world's electrical power is generated using coal, with many countries accounting for a much higher percentage: Poland gets 95% of its electricity from coal, South Africa 93%, Australia 77%, India 78%, China 76% and the USA 51%. About 70% of the coal extracted from mines feeds thermal power stations, 20% is turned into siderurgical coke for the steel industry and the rest goes into other industries (cement factories for instance) or is used domestically.

Figure 4.4 shows consumption by regions. It is striking to see how much coal is consumed in Asia: China alone devours as much as the rest of the world altogether, and has to resort to imports in spite of being by large the world's first producer. The same happens in the United States, which consume all of the almost 600 million toe they produce. Europe also relies on imports and consumes more than 500 million toe. It is noteworthy that almost all the African consumption is due to South Africa, practically the only developed nation on this continent.

But then, the regions of the world are unevenly populated. If we divide the energy consumed into the number of inhabitants, we get a more informative figure, the average share per person. In Fig. 4.5 we can see the consumption of coal per person in various countries and regions of the world. The inequality that appears between South America and Africa on one part (less than 0.1 toe per inhabitant), and the rest of the world on the other is severe. On the other end there is the USA, with almost 2 t per inhabitant. The most industrialized countries (Germany, Japan, United Kingdom) consume about 1 toe, the same as China and Russia (BP 2008).

The only exception among the big economies is France, with only 0.2 toe per person. This low figure is due to the fact that France gets almost 80% of its electricity from nuclear power. In contrast, Italy, where nuclear energy is banned by Law, so most of the electricity is generated from natural gas, all of which has to be imported, is one of the European countries with the highest energy dependency.

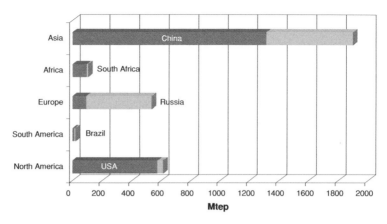

Fig. 4.4 Coal consumption per region, in tons of oil equivalent (toe). Next to each region, the main consumer is given. *Source* (BP 2008)

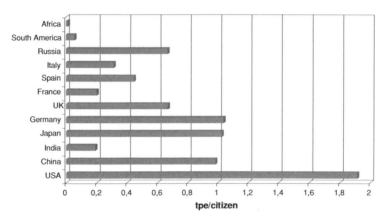

Fig. 4.5 Coal consumption per region, in tons of oil equivalent (toe). Next to each region, the main consumer is given. *Source* (BP 2008)

We often hear about the need to save in order to reduce CO_2 emissions. Burning coal is the main cause of these emissions, but at the same time it is the cheapest way to generate electricity and therefore the fossil fuel of choice for developing economies. Figure 4.5 shows that the USA is the great squanderer, but countries like Germany, Russia, Japan or the United Kingdom, who have heavily invested in alternative energies (both nuclear and, especially in recent times, renewable) still consume almost as much as China, where most electricity comes from coal.

On the African continent, excluding South Africa, there are 800 million inhabitants whose share is literally nil. Compounding them with 400 million people in South and Central America plus 1,500 million more in India, Bangladesh and Pakistan, we get 2,700 million people whose consumption of coal (and hence

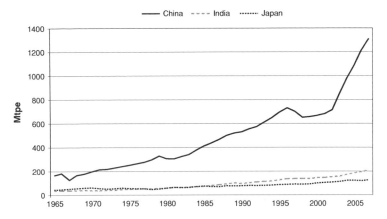

Fig. 4.6 Coal consumption in China, India and Japan

electricity) is extremely low. If we allotted 0.5 toe to each of them (as much as is consumed in Spain, half as much as in Germany, a fourth of USA consumption), coal consumption (and emissions) would *increase* by more that 1,300 million toe. On the other hand, if the figure for the USA was halved, we would save only 150 million toe. These calculations give us an idea of how severe a problem the world is facing. The conundrum can be phrased like this:

> Unless we find a way to generate electricity that is as productive as coal *but cheaper*, CO_2 emissions will continue rising as poor countries keep developing.

If you are not completely convinced, take a look at Fig. 4.6, where the evolution of coal consumption since 1965 is shown for China, India and Japan. The curve for China climbs steeply and from 2000 on it is hair rising. This frenetic development and gargantuan hunger for energy mirrors what happened during industrial revolutions, first in England and later in the USA, Japan and the European countries. It seems obvious that India, Pakistan, Brazil and other emerging economies will sooner or later follow China's example.

The Ignoble Fuel?

In the last decades, the reputation of coal has increasingly worsened, becoming the most "ignoble" of fossil fuels. But we should not forget that from the 17th century it helped avoid a first order energy crisis—biomass, which had been used until then, was being depleted—and what's more, it possibly saved the remaining European forests, which would have disappeared in no time if the new fuel had not replaced wood.

In our contemporary world, steelworks, cement factories and electricity generation depend wholly or to a great extent on coal, which, as we have seen, has a lot of important technological applications as well.

Besides, we have huge coal reserves, of which 30% are found in the USA, 17% in Russia, 13% in China, 10% in India and 9% in Australia. The "proven reserves" (that is, the amount of coal that can be mined from known deposits at a competitive price) are enormous, around one billion tons, and the so-called "total gross reserves" (proven reserves plus other fields that have not been discovered yet) are presumably much larger still. To sum up: there's enough coal for several centuries.

Electricity is essential for the countries' development, and coal is an economical means to obtain it. The great European powers depended on this mineral until quite recently; here it has been progressively replaced by natural gas in the last few decades, but for developing countries like India and China coal remains the obvious choice.

And yet, there are numerous drawbacks. To begin with, mining activities are harmful for the environment, especially when it comes to surface mining, which accounts for 60% of present day extraction. And they are hazardous. No country has been spared from tragic accidents, which in China alone cost thousands of lives a year.

A large coal-fired thermal power station provides around 1,000 MW, satisfying the needs of one million people. In exchange, it consumes 3 million tons of coal a year (equaling the global production of coal at the end of the 17th century) and releases 11 million tons of CO_2 into the atmosphere. Besides, depending on the design of the power plant and the quality of the coal, a variable number of pollutants are emitted, ranging from sulfur dioxide (the cause of acid rain) to tiny ash particles that may cause respiratory problems.

Some of these troubles can be solved, and in fact we already have commercial solutions for them. *Fluidized bed combustion* is a technology that allows to capture the sulfur and most of the ashes in the boiler, so they aren't released into the atmosphere.

The problem posed by CO_2 emissions is more difficult to solve. A lot of R&D is going on, trying to tackle it with techniques like capturing carbon dioxide in underground deposits or recycling it to get substances like methanol, which could itself be used as a fuel instead of oil. None of these approaches is commercially viable at the moment. Besides, implementing them would lead to an important rise in the production costs of electricity. Developed countries may afford to pay the extra bill, but it's unlikely that the Chinese and the Indian will.

Burning coal to obtain electricity is one of the worst pitfalls (only comparable to the trap of burning oil to move around), into which modern humans have gotten, attracted by their inexhaustible appetite for energy. It won't be easy to escape from it, but if we don't find the solution, Nature will, and, as James Lovelock keeps reminding us, irate Gaia's remedy will not be to our liking.

References

Menéndez, J. A. (2008). El carbón mineral. http://www.oviedo.es/per-sonales/carbon/carbonmineral /carbon%20mineral.htm.
Boyle, G. (2003). *Energy systems and sustainability*. NY: Oxford Press..
BP. (2008). BP world statistics. http://www.bp.com/.

Chapter 5
Manna Springing from the Earth

Grandchild: Is it true you burnt them? Did you burn all these
wonderful organic molecules?
Grandchild: It's true. I'm sorry. We burnt them.

K. S. Deffeyes

The Wilderness Experience

At one of my aunt's, there was an illustrated children's edition of the Bible. I was six or seven years old, and on Saturday afternoons she would often take care of me. It was a fortunate deal everybody gained from. My parents were free to do the shopping and get some fresh air. My aunt enjoyed stuffing me with biscuits and milk, but not as much as I enjoyed the adventures of the Bible lands: Yahweh furiously unleashing plagues on the Pharaoh, the miraculous escape from Egypt, with the Red Sea saving the tribes, in extremis, from Ramses' troops; the great king on his knees, watching the bodies of his drowned soldiers, wondering why God might favor a bunch of goat keepers; Moses wandering through the desert carrying his ark, bound for the Promised Land. And manna would drop from the sky in the early morning and seemed to me more miraculous, being so sustained and regular, than the opening of the sea.

In 1859 Colonel Drake[1] started some drillings in Titusville, Pennsylvania, which led to the first industrial oil extraction plant. Until then, people had just gathered this oily liquid, known since the time the people of Israel walked through the desert, wherever it was found. Marco Polo, who traveled to Asia at the end of the 12th century, takes note of some natural oil sources in Baku (Caspian Sea); the locals held oil in high esteem, as it provided light and heat and burnt easily. It was another kind of manna, which did not drop from the sky but sprang out of the earth, as miraculous and plentiful as the one that had nourished the tribes. Near Sinai, under the sands of the Arabian desert, there were oceans full of it.

---

[1] Edwin Drake didn't receive his chevrons from any military academy. His title was invented by *Seneca Oil,* the company he worked for, to increase his standing among the population of Titusville, where he drilled for oil.

J. J. Gómez Cadenas, *The Nuclear Environmentalist,*
DOI: 10.1007/978-88-470-2478-6_5, © Juan José Gómez Cadenas 2012

How Petroleum and Natural Gas were Formed

These oceans of crude oil, frequently found together with equally vast oceans of natural gas, were mostly formed during the Jurassic period, around 200 million years ago, when dinosaurs roamed the Earth. This gave rise to the legend that oil comes from the decomposition of the great reptiles. When I was a child, there were illustrated science books where you could see uncountable diplodocus in an underground sea of oil which seemed to flow out of them. It is true that the origin of fossil fuels is organic,[2] though less spectacular.

The process is almost identical to the way coal is formed. In certain natural environments (typically lakes, deltas and sea beds) millions of tons of remains from animals and plants settle over geological times, and are decomposed in the absence of oxygen. As time passes, innumerable layers of mud and sand heap on the organic matter, and under this pressure a material called *kerogen*, similar to peat, is formed.

Pressure rises and so does temperature. If the deposits are buried at a depth between 2 and 6 km, temperature increases from 60 to 150°C. At these temperatures, kerogen "matures" into oil or natural gas, by means of chemical reactions that break its large organic molecules into smaller pieces. Liquid oil corresponds to molecules of 5–20 carbon atoms; natural gas, to molecules with fewer than five carbon atoms and often just one, in the case of CH_4, methane, the most important component of natural gas.

Once oil and natural gas have formed, we need conditions to retain them. Oil isn't soluble in water and is lighter than it, so it tends to float on the surface, where it eventually disperses. Nevertheless, 10% of the oil gets trapped underground. This happens when an impermeable rock acts as a seal, preventing it from migrating to the surface. But even if there is such a cap rock, the oil ends up escaping on the sides, unless there is a geological trap where a *reservoir* forms.

Finally, the rock that fills up the reservoir (5 km underground, there are no hollow caves) has to be porous enough for hydrocarbons to circulate. The more porous the rock, the more oil accumulates in the reservoir.

The likelihood that all conditions are met so that crude oil reservoirs are formed and kept is very low. This is not the case with metal ores, which our planet is rich in; oil fields, however, can be considered like prizes in a most difficult geological lottery. A few places on Earth, such as Saudi Arabia, have won the jackpot. Most countries have had no luck at all.

---

[2] At least that's what most geophysicists believe, but not all of them. There might be great deposits of natural gas produced after the crash of a large meteorite.

The Black Gold

From 1920 on, due to the explosion of the automobile industry and the expansion of motorized transport, oil quickly became the most sought after fossil fuel (35% of the total primary energy in 2007), with coal (29%) and natural gas (24%) as runners-up. The growth in demand has gone hand in hand with the development of the industry of oil derivatives.

Petroleum is nothing but a mixture of complex hydrocarbons, and its various components have different physical and chemical properties, differing in their boiling points. The oldest technique used to refine petroleum is distillation, where each component (starting with the one with the lowest boiling point) is left to evaporate, keeping it apart from the rest and condensing it in a separate container. Present day refineries produce a great variety of substances derived from crude oil; other than gasoline, diesel fuel and kerosene, you get unsaturated hydrocarbons from which lubricants, solvents, detergents, wax, pharmaceutical products, insecticide, weed-killers, plastics and synthetic fibers, among many others, are obtained. We wouldn't know how to survive without them. About 6% of all the oil extracted is for the petrochemical industry, 20% for other industries and almost 60% for the most important use of *black gold*: transport.

Since Drake's discovery, an intense search for new oil fields has been going on all over the world. The first explorations were in the USA and yielded great amounts, the same in Russia, in the Baku region, at the Caspian Sea, and more recently in Siberia. In Sumatra, Java and Borneo oilfields where discovered in the 19th century. 1910 saw the beginning of explorations in the Middle East. Al-Burgan, the world's second largest oil field was found in Kuwait in 1938; number one is Al-Ghawar, in Saudi Arabia, discovered in 1948.

Nowadays, oil prospecting requires the application of sophisticated geophysical techniques, like the seismic method, based on recording the time it takes for sound waves to travel from an emitter at the Earth surface (or on a boat on the surface of the sea) to the deep rocks, get reflected by the rock crust and travel back to the surface. This technique, similar to the sonar, allows to draw geological maps of the underground formations, from where it is possible to deduce the presence of geological traps containing petrol.

Figure 5.1 shows the typical production profile of an oil field. The first step, after prospection has yielded positive results, consists in drilling a "discovery well", to make sure that there is in fact oil to be found in the region. The next well is called "evaluation well" and is used to assess how much oil can be mined from the field. Then comes the first industrial well, and from then on the production of crude oil increases as more and more wells are drilled, until we get to a plateau where production remains constant (this situation corresponds to the moment when all oil wells are operating at full production and the amount of petroleum has not yet begun to decline noticeably). Finally, after the plateau comes a decline, more or less steep depending on the sort of oil field and the extraction techniques employed. There comes a moment when it is no longer economic to mine the remaining oil and the field is abandoned.

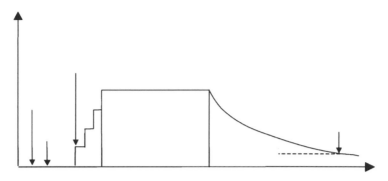

Fig. 5.1 Profile of an oilfield. *Source* (Robelius 2007)

The first oil fields where discovered on dry land, but as these got exhausted and the necessary techniques were developed, explorations were increasingly done in the sea, including the deep sea, at more than 2,000 m.

On land, oil is transported by means of giant pipelines and container trains or trucks. On the sea, the job is done by giant oil tankers, the most colossal vessels that have ever sailed, larger even than the biggest aircraft carriers. The greatest of them all, the *Seawise Giant*, weighs slightly more than 564,000 tons (ten times as much as the Titanic) and measures almost 500 m. Some of the most notorious accidents because of their environmental impact have been related to these ships (for instance *Exxon Valdez*).

American Graffiti

In the early 70s my home land, Spain, was a rapidly developing country, and everybody was working double shifts. The squalor of the postwar years was already fading, but for the ones like me, coming to age in that decade, the US was still a fabled land, where kids like us could drive those magnificent cars, racing against each other, spinning the monster engines between two red lights, with their hearts broken (who is not heartbroken, at seventeen) and their heads full of gasoline.

American Graffiti, the master work of George Lucas, was filmed in 1972, but the story is set a decade earlier. In the 60s the US industry was producing the famous Muscle Cars, monsters made of chrome and stainless steel, with 350 HP engines and fuel consumptions of fifteen or more liters per 100 km. The story offers what the title promises—graffiti, sketch, almost haiku. Four kids share the last night of their teenage years, before drifting in a different direction each. The story focuses on an illegal car race, in the small hours before dawn, on a desert road. We, too, knew about illegal races on desert roads in our provincial Spain, except that we were riding small, engine-doctored motorcycles, while they drove the unforgettable Cadillac el Dorado, Pontiac Firebird and Ford Mustang. We looked at them, and the smoke of those chromed exhausts seemed to form mirages of a Paradise of freedom and excess.

In 1972 the price of oil was 3$ a barrel. At the end of 1974, the price had increased fourfold, after the OPEC[3] countries, led by Saudi Arabia, imposed an oil embargo on the USA and several other Western countries as an answer to their support of Israel in the Yom Kippur war.

The embargo resulted in a crisis that was as severe as it was unexpected. In the USA, the price of gasoline rose by more than 50% in less than a year, and the New York stock exchange plummeted, with losses of 97 billion dollars in 6 weeks. The sudden shortage of fuel caused inflation, economic recession and unemployment, with thousands of lost jobs.

Standing in long queues to fill up the gas tank or, still worse, finding that the gas station at the corner is closed for lack of supply, drove Americans to something like a collective nervous breakdown. All of sudden, the end of the world seemed inevitable. The philosopher E.F. Schumacher summarized the spirit of the times in a catchy phrase that might as well describe the current moment: *the party is over.*[4]

The USA weren't the only country affected by OPEC's retaliation. Portugal, Denmark, Netherlands, United Kingdom and Japan had to tighten their belts as well. In Spain, the crisis stopped an enormous growth of around 7% yearly that the country had been experiencing since 1961. Construction and tourism, the two driving forces then and now, suffered directly from the shortage of oil, which also had an effect on the textile, shipbuilding and automobile industry. My generation spent their teenage years in a country burdened by unemployment and inflation, where political transition veiled, up to a certain degree, a stagnant economy we wouldn't leave behind until well into the 80s.

But the worst was yet to come, with the fall of the Shah of Iran and the Islamic revolution that temporarily halted oil production in one of the world's main exporting countries. In 1974 the price of a barrel had increased fourfold, but in 1980 it was multiplied by ten (Fig. 5.2). More mass hysteria and more restrictive measures in the USA, where schools were closed to save on heating, speed was regulated on all motorways and rationing coupons were ready to be distributed, though this did not come into force.

In 1979, Jimmy Carter had solar panels installed on the roof of the White House. But the USA didn't just go for saving and considering the use of alternative energies. In January 1980 Carter declared that any interference with US oil interests in the Persian Gulf would be deemed an attack on the vital interests of his country and therefore "morally equivalent to a declaration of war". With this statement, the president of the most powerful nation on Earth expressed the possibility to intervene militarily in sovereign countries in order to ensure the provision of oil, which in fact happened 10 years later in Iraq.

---

[3] Organization of the Petrol Exporting Countries. Its founding members are Saudi Arabia, Iraq, Iran, Kuwait and Venezuela.

[4] Two decades later, R. Heinberg would use this phrase as the title of a book about the oil shortage (Heinberg 2005).

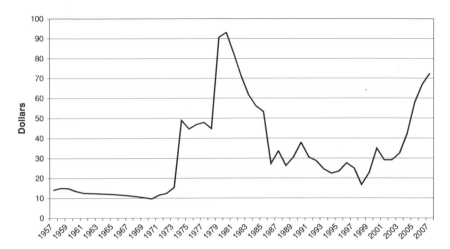

Fig. 5.2 Price of an oil barrel from 1957 to 2007. The steep peak corresponding to the energy crisis in the 70s is clearly visible. *Source* (BP 2008)

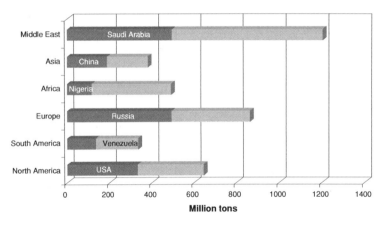

Fig. 5.3 Oil production per region. Next to each region, the main producer is given. *Source* (BP 2008)

The 80s were the decade when consumption went down. The sudden stop imposed by OPEC halted the prodigious, or insane, growth of half a century. The landscape of Fig. 5.3 shows a steep peak in 1979, followed by years of decline. When production starts to rise again, from the mid 80s, it does so much more gradually. The West had learned the lesson in austerity and for the first time demand was below supply. The market took care of the rest. OPEC was forced to release the prey and prices dropped until the end of the 90s, settling around pre 1973 levels.

With lower petroleum prices and economic recovery came amnesia. Most of the energy saving measures introduced in the 70s were abandoned, the drive to develop alternative energies was halted, as they were not economically competitive. Muscle cars enjoyed a comeback, disguised as Jeep Cherokee, Range Rover, Nissan Patrol, BMW X5. With cheap kerosene, low cost airlines become profitable, and European get used to shopping in London, a romantic weekend in Amsterdam, or quick cultural tourism in Paris, Madrid or Rome. The automobile industry thrives, ushering in traffic jams, pollution, three thousand fatal victims per year in traffic accidents in Spain alone. The solar panels at the White House were removed by the Reagan administration in the mid 80s, a decision as emblematic and deliberate—albeit pointing in the opposite direction—as the decision to install them.

Oil in the 21st Century

3.9 billion tons of oil was mined in the year 2007, at a crazy rate of 81 million barrels a day. Figure 5.5 shows the distribution per region. The countries of the Persian Golf are the leaders of the pack, with the mind-boggling figure of 1.2 billion tons (31% of the total production, from which 12.5% correspond to Saudi Arabia), followed by Europe (22% of the total, half of which is due to Russia) and the North American region (15% in total, with 9% from the USA).

The fact that oil production is in the hands of only a few countries has been, since the energy crises in the 70s, an unremitting source of problems and instability. OPEC controls 43% of the global oil production in 2007 and hoards 62% of the known reserves. Together with Russia, this gang of countries is capable of putting the world in check by rising prices or limiting production. And it's a dangerous game, as has become clear with the two Iraq invasions led by the USA and aptly named oil wars.

Figure 5.4 allows us to understand—which does not mean justify—American aggressivity when it comes to oil. This country alone gobbles up a quarter of all the petroleum mined worldwide. And yet, the Asiatic region already consumes more oil than North America, given the demand of emerging economies, headed by China, including also India, Indonesia, Thailand and Singapore, among others. European countries throw in another 25%. These three regions consume almost 85% of the oil extracted from fields most of which lie somewhere else, that is, in OPEC countries or in Russia.

Spain is a typical case. The country produces a meager 140,000 tons, compared with almost 70 million tons consumed. Of the coal consumed, 70% is imported, with oil it's practically 100%. Spain imports from the Persian Gulf (24%), Europe (30%, half of it from Russia), Africa (27%). The remaining 20% comes form Mexico, Venezuela and other countries. To express it bluntly, Spaniards are at the mercy of OPEC's and Russia's whims.

Petroleum has been getting more expensive since the beginning of the Millennium. Halfway through 2008 the price reached the historical record of 150$ a barrel and then plummeted again under 50$. However, one of the few things all

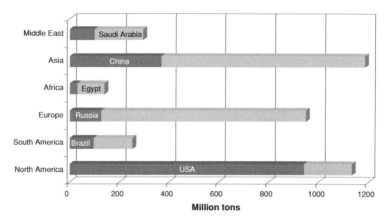

Fig. 5.4 Oil consumption per region. Next to each region, the main consumer is given. *Source* (BP 2008)

economists and international analysts seem to agree on is that these low prices will not last.

Why is oil so important for the economy? The answer is obvious once you come to realize that every piece of merchandise and every passenger that is transported by land, air or sea from one point to another point on the globe consumes petrol, diesel oil or kerosene, and the same goes for every farming machine and every fishing boat. And that's without counting plastics, drugs, pesticides und numerous other derivative products. The world quite literally depends on those 81 million barrels a day in order to keep turning. If the price of oil goes up, the price of essential goods like food goes up too, for two reasons: production costs rise (the diesel oil consumed by fishing boats and tractors gets more expensive, and so do potatoes and herrings), and more has to be paid for transport. From the taxicab to the airplane, all services are affected. People find it hard to get to the end of the month and so on.

The question that springs to mind is, how much is left? One thing we all know for sure is that the manna isn't going to last forever.

The Problematic Question of Oil Reserves

The question is not easy to answer for a lot of reasons. To begin with, we don't know exactly how many oil fields remain to be discovered. Very few, according to the most pessimistic (Campbell and Laherrère 1988); as many as we have found up to know, say the optimists (USGS 2000). What's more, we don't know precisely how much oil we can *recover* from these fields. As we said before, a well is not abandoned when it has "dried out", but when it is no longer profitable to continue exploiting it. On the other hand, the use of innovative techniques—such as

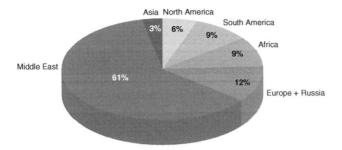

Fig. 5.5 Distribution of oil reserves by region. *Source* (BP 2008)

transversal drilling, which allows to open side wells from the main well, thus increasing the amount of oil you can get from a well—may become profitable in the future. Similarly, if the price of oil is very high, companies are more interested in squeezing out the wells up to the last drop than if the price is low and the cost of extracting the last ten or twenty percent exceeds the benefit. As a result, the "proven reserves" (that is, the amount of oil that is estimated can be recovered economically) are hard to assess. To make matters worse, then there's politics. In some countries, such as the USA, information about oil reserves is in the public domain. In others, which turn out to be the greatest producers (Saudi Arabia, Russia, Iran) this is classified information and not all experts trust the figures that are made public.

The data about oil reserves are even more alarming than the facts related to production. The countries in the Persian Gulf hoard 62% of the reserves (Fig. 5.5). If we calculate for how many years they would last *given the current consumption rate* we find that the Middle East could satisfy the demands of the whole world for slightly more than a quarter of a century, while no other region has enough to last for 5 years. Besides, mining is cheap in these countries: it costs under 2$ a barrel, compared with extraction costs of 8–10$ a barrel in the North Sea (IEA 2002).

Saudi Arabia is not only the greatest oil producer, but also holds the greatest reserves (36 billion tons, 21.3% of the world reserves, enough to supply the planet for 10 years). The largest oil field ever discovered is in this country: Ghawar contains about 7% of the known reserves on Earth. Here there are 60 oil wells, though only a few of them are being exploited. In contrast, they consume less than a 100 million tons, just 20% more than for example France or Spain, countries that for all practical purposes have no petroleum.

How Much Oil Remains to be Discovered?

Might there be another Ghawar under the Siberian tundra, the Antarctic ice caps or the sands of some distant sea? Opinions differ.

For a great number of experts (Deffeyes 2005; Campbell and Laherrère 1988) there isn't much more to be found. They argue that the number of oil fields has been going down in the last 50 years. These authors also think that the total existing reserves have probably been exaggerated.[5] According to these viewpoints, the total amount of recoverable oil on the planet is approximately two hundred billion tons, of which about one half has been extracted.

In contrast, one of the most important geological studies carried out in recent times, the U.S. Geological Survey (USGS 2000) reckons that yet undiscovered oil amounts to a 100 billion tons, as much as has been extracted. On top of it, they anticipate that the existing reserves (160 billion tons) will reach 200 billion in the next decades (owing to improved mining techniques and higher prices which will allow to exploit up to the last drop of those oil fields which are today abandoned as unprofitable). On the whole, the USGS gives a figure of 400 billion tons for the total recoverable oil, of which 25% has been extracted.

Finally, we have to take into account the so-called non-conventional oil reserves. Let's mention only the most important. In Canada and Venezuela there are large deposits of tar sands. They are created when an oil field rises to the surface after being formed, so that the lighter hydrocarbons evaporate, leaving behind an almost solid residue. In a way they are failed oil fields, a consolation prize in the geological lottery that advertises conventional oil fields. In fact, the regions of Athabaska and Cold Lake, in Canada, once held ten times as much oil as Ghawar. The same happens with the tar sands in Orinoco and Maracaibo, in Venezuela.

The U.S. Geological Survey (USGS 2000) considers that around 10% of these huge quantities of tar can be extracted with present day techniques, so Canada and Venezuela would each of them possess reserves the size of those held by Saudi Arabia. Not bad for a consolation prize, as long as it is really feasible to obtain oil at reasonable prices in the future. The technique employed in the Athabaska deposits consists in injecting hot high-pressure steam in order to heat the tar to liquefaction point. For this you need immense amounts of water and natural gas, resources which are themselves limited. Another problem associated with exploiting the tar sands is environmental impact. And even so, the price of non-conventional Canadian petroleum is starting to be competitive, at around 10$ the barrel, not much more expensive than North Sea oil.

Twenty Years are Nothing

The production of oil is uncertain, and then there is concern related to consumption. On one side there's the North American giant, devouring a quarter of all the world's petroleum, including its own still copious production and growing

[5] Let's not forget that reserves can increase without new oil fields being discovered, if it's considered possible to extract more oil from existing fields, at a profit.

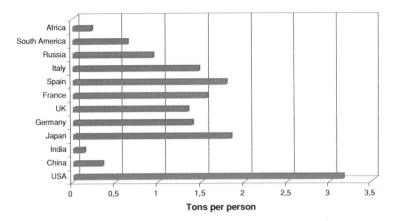

Fig. 5.6 Per capita oil consumption in several countries around the worlds. *Source* (BP 2008)

imports. Its 300 million inhabitants consume 943 tons per year, that is, a bit more than 3 tons per capita. On the other side there's the European countries and Japan, who consume around 1.5 tons per capita, half as much as the USA (Fig. 5.6). And China got by with the oil it produced until the year 1995, when both the consumption of oil and of coal boosted; today it stands at 350 million tons.

Was it really a boost? China has 1.3 billion inhabitants, this means their per capita consumption in 2007 was a measly 0.27 tons per capita. Consumption is even lower in India and in all of Africa, except South Africa.

What happens when a country with more than a billion inhabitants becomes industrialized to levels similar to the developed world? The Chinese not only ambition electricity, heating and household appliances in their homes—which has turned them into gargantuan consumers of coal—but they also aspire to drive their own motor vehicles, something that is still out of reach for most of the population. To deny them this right while in the West we rely ever more on our cars is pure hypocrisy.[6] To accept it makes our hairs stand on end, especially if we add Indians and Pakistanis to the equation. If the pessimistic forecasts are correct, then there's only enough oil for 40 years at current consumption rates. But if we assume that the inhabitants of the emerging economies (more than 2.5 billion) have the right to consume just one ton per person (50% less than Europeans, only a third of USA levels), consumption would increase by two billion tons a year. Adding 800 million Africans we would almost have to double the crazy rate at which oil is mined, and then reserves would only last for two decades. So there are 20 years to go. And 20 years, as the old tango says, are nothing.

---

[6] Of which we are perfectly capable. Nor are we upset by the millions of avoidable deaths per year, caused by typhus, malaria, undernourishment and poor sanitary conditions all over Africa. But we are distressed by the fact that climate change may cause the extinction of polar bears.

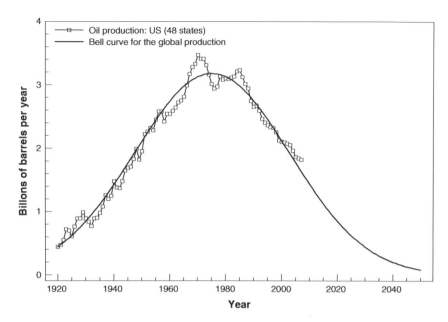

Fig. 5.7 Oil production in the USA from 1920 to 2007 [Source (IEA 2008)], with the Bell curve predicted by Hubbert's theory

Is the Party Over?

What would happen if there was a new crisis, not due to "political" reasons—if OPEC's manipulations are worthy to be called political—but to a global shortage of oil? This is the subject of plenty of recent books, some of them sensationalist, like "The party is over" (Heinberg 2005), whose central thesis is that the growing gap between an increasing demand and a dwindling production will lead to a global crisis with devastating consequences.

By the way, Heinberg doesn't say anything new.[7] Since the 70s crisis, there have been a lot of more or less rigorous analyses. The common denominator of all of them is the hypothesis that the global oil production—as the natural gas production and eventually the coal production—follows a bell-shaped Gauss curve: it increases at the beginning, reaches a maximum and from then on it decreases at the same rate it had grown before reaching the maximum. This hypothesis is known as the peak oil theory, or Hubbert's theory, after the American geophysicist Marion King Hubbert (1903–1989), who correctly predicted (Hubbert 1956) that the total oil production in the USA would reach its peak towards the end of the 60s/ beginning of the 70s and then start to decline. Figure 5.7 shows that US oil

---

[7] Not even in the title of his book (The Party is Over), which he borrows from philosopher Schumacher.

production can in fact be approximated by a Gauss curve. Hubbert's theory states that the world's oil production will follow the same pattern.

The peak of the Gauss bell corresponds to the point where the curve has swept half of the surface (it's symmetrical, the right part of the bell is a mirror image of the left part). The total surface corresponds to the amount of recoverable oil on the planet. According to the most pessimistic authors, we have already consumed almost half of the recoverable total, so we are right at the peak of the bell.

What happens then? When the peak is reached in a specific region, like the USA, petrol can be imported to make up for the growing difference between demand and production (that is just what is happening in North America). But where do we import oil from when the global peak is reached?

The pessimistic answer is as follows: as time goes by, the difference between demand and production keeps rising. Oil prices skyrocket, and so does inflation. Shortage increases and affects transport, so there are no supplies in supermarkets and long queues at the petrol stations. Mass hysteria, restrictions, chaos like in 1972 and 1979, but now *seriously*.

For some authors, like Heinberg, the end of industrial civilization is unavoidable. The shortage of oil and its uneven distribution will lead to new "oil wars" (which we are already familiar with after two wars in Iraq). The towns, with no food due to failing transport, and no electricity due to power outages, will become mousetraps. The rich will fence themselves off, surrounded by their bodyguards, weapons and the few oilcans left. In brief: Mad Max.

The equivalent of this crisis in the 17th century, pictured by the Heinbergs of the time, would have been a devastated Earth with not a single tree standing, and survivors of the massacres roaming around, frozen and starving. It's true that this never happened thanks to coal. The most optimistic experts think that we are still far from having used up half of the oil on the planet, that we still hold 80% of the capital, maybe much more taking into account unconventional sources, so that in any case we would have time enough to develop alternatives.

In my opinion, the staunch supporters of the peak oil theory exaggerate both its imminence and its possible consequences. In contrast with electricity, which our society can't go without for even an instant, oil is quite an elastic resource, as was proven by the great 70s crisis and then the 80s depression; both events were, in a way, a rehearsal for the advent—or rather the non advent—of the renowned oil peak. When the OPEC artificially limited their production, consumption started to fall after an interval of panic and confusion, and so did prices and inflation. Something similar is likely to happen when oil reserves start to get scarce. It will probably result in—one more—crisis, but the increase in oil prices will lead, on the one hand, to moderate consumption, and on the other will provide incentives for exploration and better use of unconventional resources.

But then, arguments by experts like Deffeyes (2005), Cambell or Laherrère (1988) can't be dismissed lightly, and suggest that in any case the end of cheap oil is just around the corner. We have to understand that the global consumption is so enormous—and, what's worse, keeps growing at an ever faster rate, with China, India and other emerging economies joining the feast—that doubling the existing reserves would just allow us to buy a couple of decades more. It seems unavoidable that the more oil we have, the more we gobble up, wasting any surplus in no time at all.

Up to the moment, catastrophist prophecies have turned out to be wrong and current oil prices, under 50$ the barrel (as of beginning of 2009) are, curiously enough, one of the few good news in the middle of the global crisis. But we mustn't fool ourselves. The transition from biomass to coal took more than a century to be completed, and there's no reason to think that the transition from oil to an alternative that doesn't exist yet, except in our imagination (like the famous hydrogen economy) will happen quickly enough to satisfy our needs. Perhaps we shouldn't forget that the manna isn't going to last forever.

References

BP. (2008). *BP world statistics*. http://www.bp.com/.

Campbell, C. J. & Laherrère J. H. (1988). «The end of cheap oil», Scientific American, March.

Deffeyes, K. S. (2005). *Beyond oil*. USA: Hill and Wang.

Heinberg, R. (2005). *The party is over*. USA: New Society.

Hubbert, M. K. (1956). *Nuclear energy and the fossil fuels*. San Antonio, Texas: Spring meeting of the American Petroleum Institute.

IEA. (International Energy Agency, 2002). *World energy outlook*. http://www.iea.org/Textbase/nppdf/free/2000/weo2002.pdf.

IEA. (International Energy Agency, 2008). *World energy outlook*. http://www.worldenergyoutlook.org/2008.asp.

Robelius, F. (2007). *Giant oil fields and their importance for future oil production*. PhD thesis, Faculty of Science and Technology, University of Uppsala, Uppsala.

USGS. (2000). *U.S. Geological survey world petroleum assessment, 2000. Description and Results*. http://pubs.usgs.gov/dds/dds-60.

Chapter 6
The Sacred Fire

According to the legend, a goatherd found a sacred fire rising out of the earth. Around this fire a temple was built, and the woman living there could foretell the future by looking into the flames. She was called the Oracle of Delphi.

Fen Fires

The fire on Mount Parnassus that gave rise to the temple of the Oracle of Delphi was not the only spontaneous outflow of natural gas known to the Ancient World. People in ancient India, Persia and China knew about these fen fires, and there they also used to attribute them to supernatural causes, at least until 500 BC, when the Chinese built the first gas pipelines—using bamboo!—many centuries ahead of the West, as on many other occasions.

In the USA, the existence of natural sources of gas has been known since 1626, when the French exploring the region of Lake Erie watched the native tribes lighting fires at outlets around the lake. In fact, the well drilled by Drake in 1859 provided not just oil but also natural gas, which was transported to Titusville, 9 km from there, through a gas pipeline. Though very simple, this first installation proved that gas could be exploited profitably. Edwin Drake's military credentials may have been false, but undeniably he initiated both the oil age and the natural gas age.

A few years later, in 1885, Robert Bunsen invented his famous burner (Fig. 6.1), a device that mixes natural gas and air in the appropriate proportions, thus obtaining a very hot, blue low-light flame that can be regulated (mixing in more or less air) and is safe to use for cooking and heating; all modern gas appliances are based on it.

Natural gas is, in many ways, a better fuel than coal or oil, but its transportation over long distances poses serious difficulties. The most economical means is by way of gas pipelines, made from steel or carbon fiber, that carry the compressed gas over thousands of kilometers. This is the way gas is transported in Spain, whose national net of gas pipelines is shown in Fig. 6.1. One of the main branches comes directly from Algeria, across the straight of Gibraltar; submarine pipelines are feasible if it's just a few kilometers. The other branches reach out from harbor cities where gas arrives in liquid form.

J. J. Gómez Cadenas, *The Nuclear Environmentalist*,
DOI: 10.1007/978-88-470-2478-6_6, © Juan José Gómez Cadenas 2012

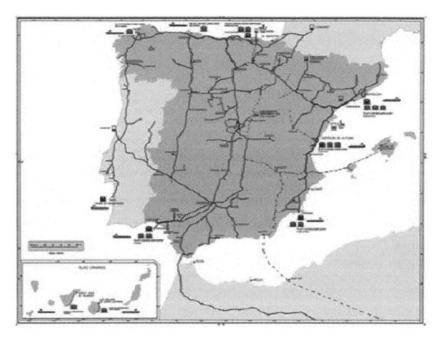

Fig. 6.1 Map showing the gas pipelines in Spain. *Source* (Foro Nuclear 2008)

Large-scale construction of gas pipelines did not begin until 1920, and their development took four or five decades. However, intercontinental transport is waterborne, using great boats like the one shown in Fig. 6.2, similar to oil tankers. Natural gas has to be previously converted into LNG (Liquid Natural Gas), as it would otherwise be too bulky. In the destination harbor, large regasifying plants are built, connected to the pipeline network.

The liquefaction process is expensive from the energy point of view, so it increases the price of the resource, on top of the costly boats that carry the gas in enormous cryogenic tanks at a temperature of not more than − 162°C (the boiling point of natural gas).

The process is not without risks. In 1944, an explosion at a gas storage tank killed 128 people and left many more injured in Cleveland, Ohio. And in 2003, 27 people died when a liquefaction plant exploded in Algeria. A recent study carried out by the national laboratory Sandia in the USA (Hightower et al. 2004) concluded that an explosion due to a leak in a ship carrying LNG could lead to a major disaster. Apart from the accident hazard, these boats are an obvious target for terrorists.

Sounds familiar? These are the old arguments against nuclear energy (risk of serious accidents, based on past catastrophes, difficulty of transportation, risk of terrorist attacks), except that in the case of natural gas, public attention is not usually focused on these disasters, which are as real as catastrophes in coal mines,

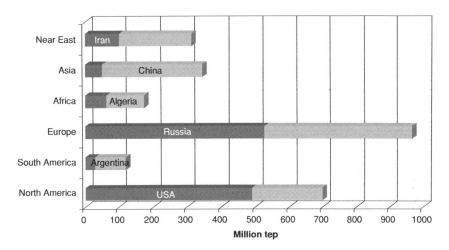

Fig. 6.2 World production of natural gas, by regions, with the largest producer. *Source* (BP 2008)

fires in oil wells, shipwrecks of super tankers and many more calamities related with obtaining and transportation of fossil fuels. Some of the accidents could be avoided with better safety measures that should certainly be demanded from governments and companies, like in fact it has happened with nuclear power stations, one of the most strictly regulated and controlled industrial activities in all the world. But there is a fraction of risk that a society managing an enormous energy flow like ours must assume. Be it the transport of passengers or merchandise, by rail or air, be it the exploitation of fossil and other alternative fuels, we can and must demand that risks are minimized and continuously and strictly monitorized (two examples are civil aviation and the supervision of nuclear power plants). But the demand for zero risk is an illusion.

Natural Gas in the 21st Century

For the last three decades, the world demand for natural gas has increased much faster than the demand for coal or oil. This has been called the "dash for gas", and its causes are not surprising. Natural gas is a cheap, convenient fuel for household uses (cooking and heating): easy to distribute once the network is in place, clean, with a high calorific value, burning without odor, residues or pollution. The same arguments, together with the fact that it emits half as much CO_2 per energy unit as coal, make it the ideal fuel for thermal power plants, like the CCGT (Combined Cycle Gas Turbine). Thermal power stations using natural gas are cheaper to build than coal or nuclear plants, besides they are excellent for following the "peak demand". This need is stressed especially in countries where wind power is starting to play a significant role for electricity generation. As we will see in

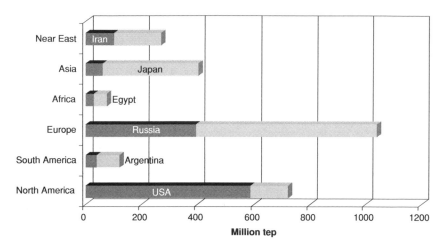

Fig. 6.3 World consumption of natural gas, by regions, with the largest consumer. *Source* (BP 2008)

Chap. 12, one of the main limitations of this energy (and, in general, of all renewables) is its intermittence, that is, the fact that the power is only available during a fraction of the time, so "operational reserves" are needed to meet the demand when there isn't enough wind. These reserves are generally CCGTs.

Production and Consumption

Figures 6.2 and 6.3 show production and consumption of natural gas in different regions. Here, the asymmetry between countries producing and countries consuming this resource is much less pronounced than in the case of oil. The USA is the greediest consumer, as usual, but they're also the second world producer, not far behind Russia, who not only heads production, but is also a voracious consumer. There's a tendency among the countries with plenty of natural gas reserves (Russia, United States) or with neighbors who are important producers (Japan, Europe) to make generous use of this resource. On the other hand, emerging economies like China or India are less interested in it—among other reasons, because of difficulties in transport and high prices—so they prefer to rely on coal for electricity generation.

Reserves

The same as with oil, natural gas reserves are concentrated in the Persian Gulf (41%), and specifically in two countries: Iran (15%) and Qatar (15%). Russia holds

25% of the total reserves, so more than half of the future natural gas production depends on these three countries, of which one is a superpower, while another has been, and still is, involved in international conflicts. On the other hand, North American reserves are small (5%), as are European ones without Russia. This implies that in the future we will see a greater dependence on a handful of producing countries, and a greater LNG marine traffic when the USA have to import it form Africa and the Persian Gulf.

The proven natural gas reserves are greater than the oil reserves, about 60% more, so the shortage of natural gas (or rather the peak that marks the start of the decline) would in principle be delayed for two or three decades with respect to oil. But then, there is a tendency in Europe and in the USA to increase dependency on natural gas, for different reasons: it is an excellent alternative to coal for electricity generation with lower CO_2 emissions, it is still cheap enough to be competitive for household usage, and CCGTs are necessary to make up for the low efficiency of wind power. On the other hand, the year 2009 has brought news reminding all of Europe that Russia holds the key that opens and closes the tap to this precious resource.

Obviously, Spain has to import almost all of the natural gas the country consumes. The Spanish strategy consists in building regasifying plants connected to the national net of gas pipelines, which allows to import gas transported by sea, mainly from Algeria, but also from countries in the Persian Gulf, Nigeria, Egypt and Libya, plus a small amount from Norway. Spain is therefore one of the few European countries that does not depend on Russian gas. And that's an excellent idea.

What about the price of natural gas? In the last decades, it has been rising continuously. This tendency will most probably not be inverted in the future, given the high concentration of reserves (a monopoly that allows a reduced group to control the price, as happens with OPEC), and the higher cost associated with sea transportation and the construction of liquefaction and regasification plants. It's not an encouraging scenario.

References

Foro Nuclear (2008). http://www.foronuclear.org.
BP (2008). BP world statistics. http://www.bp.com/.
Hightower, M., et al. (2004). *Risk analysis and safety implications of lng spill over water.* Albuquerque: Sandia National Laboratories.

Chapter 7
On Board the Nautilus

*There is a powerful agent, obedient, rapid, easy, which
conforms to every use, and reigns supreme on board my vessel.
Everything is done by means of it. It lights, warms it, and is the
soul of my mechanical apparatus. This agent is electricity.*
Jules Verne, "20,000 Leagues Under the Sea"

The Earth in Darkness

Some months ago, I took part, with some friends, in an astronomy trip to the
mountain Pico del Buitre, in Javalambre (province of Teruel, Spain). The trip,
among whose participants there were prospective scientists between the age of four
and nine, included a night devoted to amateur observation with portable
telescopes. El Pico del Buitre is one of the darkest spots in Spain —and also one of
the coldest, as all members of the party remember well. Even darker is Roque de
los Muchachos, on Las Palmas (Gran Canaria, Canary Islands), one of the most
important sites for astronomy observation on the Northern hemisphere.

There aren't many places left in Europe, and even less in North America, which
are adequate for this aim. Our planet is like a village square on a festival night.
Light pollution is something we are so used to that spending a genuinely dark night
outdoors can turn into an almost mystical experience, especially under the distant
and eternally mysterious roof of the constellations.

However, when Thomas Alva Edison (1847–1931) was born, not much more
than hundred and sixty years ago, a spaceship heading for the Earth—like those in
the novels by Isaac Asimov and Arthur Clarke—would have revealed a dark
planet, with a few gas streetlamps lighting the streets too dim to be detected.

Edison is credited for the phonograph and, in part, for the light bulb. He was
also the first inventor who applied the principles of mass production and industrial
assemblage to the development and marketing of his patents. His lab in Menlo
Park, New Jersey, is justly considered the first industrial research laboratory in the
world.

As it often happens with biographies of great men, Edison's attracts and puts us
off at the same time. It is difficult not to sympathize with the poor, almost deaf
child—possibly due to scarlet fever—who made his living selling sweets and
newspapers on the trains connecting Detroit and Port Huron. Or to admire the
explosive mixture of talent and energy he put into founding 14 companies, among
them the mythical *General Electric*. The romantic image peaks when we listen to

J. J. Gómez Cadenas, *The Nuclear Environmentalist*,
DOI: 10.1007/978-88-470-2478-6_7, © Juan José Gómez Cadenas 2012

his voice reciting "Mary had a little lamb", reproduced on the phonograph he invented. The 1927 recording renders the voice of an enthusiastic, good-natured old man. A nice voice, a bit like a farmer's, almost naïve.

Appearances are deceptive, we all know. Thomas Alva, the businessman, the avid inventor, wasn't naïve at all. In Menlo Park, his employees, engineers and technicians worked at the blast of a bugle. A great part of his patents—which were always on his name—were based on work done by other researchers. This was the case with the light bulb, of which there were previous designs, developed by inventors such as Joseph Swan and William E. Sawyer, among others. In fact, from 1883 to 1889 Edison fought legal battles with Swan over the validity of his patent, which were settled with the creation of a joint company, *Ediswan*, to manufacture and market the product in England.

It has to be admitted that, in spite of its practical and iconic value—few objects define the 20th century better than the light bulb—the invention in itself didn't guarantee the prodigious phenomenon of electrification. In order to set up the first commercial system for the generation, transmission and conversion of electricity, you need outstanding talent and fanatic stubbornness, of which our hero had more than enough. If electric power is the genie that is always ready to grant us all our wishes, Edison was the Aladdin who managed to trap it inside the lamp.

The problems he faced would have discouraged any less messianic person. To begin with, he was obliged to generate electricity at a price to compete with gas lighting, a well-established technique at the time. Edison and his company, General Electric, lacked the formidable infrastructures that were already in place in all Western cities to generate and transmit the gas that lit streets and homes. Quite literally, he had to start from zero.

It was to his advantage that electric power is clean, convenient and safe. Gas lamps used to produce plenty of heat, steam and carbonic acid, and caused frequent accidents: choking and often explosion due to leaks. But on the other hand all the elements of a yet inexistent electric network had to be designed, financed, built and put into operation, from power plants to transmission cables that had to connect the plants and the neighborhoods or industrial areas, to electrical appliances—light bulbs—unknown to most of the public. And then the system had to be reliable from the beginning. Frequent power cuts would have quenched the invention right at the start, or at least would have finished with General Electric, delaying the widespread use of electricity for who knows how many decades.

In 1881 Edison installed a complete system (generators, distribution circuits and light bulbs) in the Hinds, Ketcham & Company printing firm in New York. The electric light allowed to work in night shifts, which had been impossible up to then because gaslight doesn't allow to distinguish colors well. The year after Holborn Viaduct in London was lit (which allowed to tend the cables under it, instead of burying them in the streets or using pylons). Figure 7.1 shows one of the "jumbo" dynamos (devices generating direct current) of the company, which was probably installed in this first power station. The same year they started the installation at Pearl Street, New York, with six Jumbos, each lighting 400 bulbs.

Fig. 7.1 Edison's "Jumbo" dynamo, probably installed in Holborn, London, 1882. *Source* Edison National Historic Site

At the beginning of 1882, when Edison started to collect money for electricity in the neighborhood, his plant would light more than five hundred bulbs. In 1883 General Electric lost money because of the frequent dynamo breakdowns, but in 1884, with more that ten thousand bulbs connected to the system, they made a profit. It was the birth of one of the greatest industries in history.

Charles Parsons's Invention

It has to be admitted that the first electric plant was a monument to wasting. The giant Jumbos basically generated heat, so that just a tiny part of the energy from the burnt fuel (around 5%) was transformed into electricity. Therefore enormous amounts of coal were needed to produce only a little light.

Edison had invented electric light and had managed to conceive a system to take lighting to the cities, *but without any profound innovations allowing a higher efficiency in energy conversion, the system wasn't viable.* We will come back to this difficulty when we look at the exploitation of solar energy.

Astoundingly, this profound innovation without which electricity would have been but a luxury for rich people, appeared in the last prodigious decades of the 19th century: the steam turbine, or turbo generator, invented by Charles Algernon Parsons (1854–1931).

Charles Parsons's background is very different form Edison's. He was born into a noble Irish family, wealthy, cultured and devoted to science: his father, William Parsons, the third Earl of Rosse, was a famous astronomer. Charles studied first at

Trinity College, Dublin, and then graduated from Cambridge with a first-class honors degree in mathematics, as befitted his elevated intelligence and his elevated social rank.

Fortunately, talent doesn't discriminate among social classes. Edison transformed his likely destiny (anybody else in his circumstances would have been satisfied with a job as a telegrapher or a shop assistant), and so did young Parsons. Instead of choosing a comfortable career at university, or a life of delightful idleness, he joined an engineering firm as an apprentice. Without this unusual step, he might never have developed his invention.

This first prototype was quite rudimentary, but the brilliant idea had turned to metal. The rest is engineering. A condensator had to be added, the blades improved, thousands of moving parts had to work with absolute efficiency and safety under high pressure and temperature. There's the engine of commercial aircraft to prove the reach of this machine, one of the most important pillars (and possibly the least well known) on which the modern world stands (Fig. 7.2).

The Electric Network

When Captain Nemo addresses Pierre Aronnax to describe electricity as the genie that lights the Nautilus, we feel he's talking to us directly. Like in the marvelous submarine imagined by Verne, this ubiquitous, powerful agent is at our reach, everywhere: we just have to press a switch. It lights our homes and keeps us warm in winter and cool in summer. It gives life to the TV, the dishwasher, the washing machine, the fridge, the vacuum cleaner, the computer, the telephone, the radio. Without electricity literally nothing would work on board our computerized Noah's Ark.

At the beginning of the 1980s, in Spain there were no more than three or four large computers powerful enough for me to do the calculations needed for my PhD project. We communicated with the machine by means of a pack of punched cards. Each card contained programming instructions, and the pack consisted of hundreds and often thousands of cards we carried in shoeboxes to the computing centre, where the UNIVAC operators took care of them in exchange for a receipt and a compassionate smile. Two weeks later we got them back, together with the computer printout. When everything had gone well, we had some results and the process was repeated, but that hardly ever happened. Just a single mistake in the code written in FORTRAN (the programming language used to communicate with the machine) and two weeks work was ruined. Often the cards got lost or were introduced in the wrong order or they got torn in the machine swallowing them… and even so we thought of ourselves as privileged, having access to this supercomputer, whose outrageous price and costly maintenance—the machine took up one whole floor and had a little army of technicians and operators at its service—only an important research centre could afford. That was not more than 25 years ago. The little laptop on which I'm typing these words is thousands of

Fig. 7.2 A modern turbo generator in an electrical power station, built by Siemens

times faster and more powerful than the punched card monster. But both machines, as different from each other as a stagecoach is from a spaceship, have something in common: both need electricity to work.

In a few decades, Spain has stridden from the Middle Ages to the Information Age: the ADSL network connects millions of homes and there isn't a single activity that is not, in one way or the other, related to the Internet. To young people at least, it seems as if things only exist if Google finds them. Writing a letter on paper is a romantic anachronism; managing our bank accounts, booking hotels or

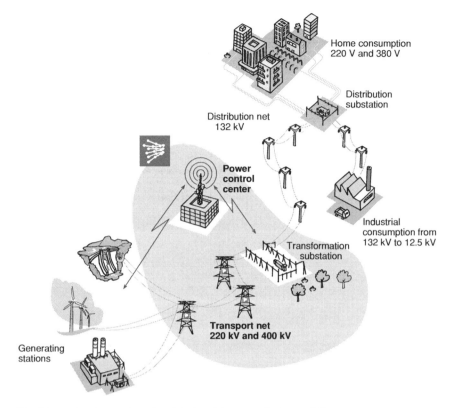

Home consumption
220 V and 380 V

Distribution
substation

Distribution net
132 kV

**Power
control
center**

Industrial
consumption from
132 kV to 12.5 kV

Transformation
substation

**Transport net
220 kV and 400 kV**

Generating
stations

Fig. 7.3 Basic scheme of the Spanish electricity network. *Source* REE

flights, the weekly shopping, medical appointments, all this is handled on the Internet, which in turn is fed by the invisible energy we call electricity.

This invisible energy is generated in thermal power plants (coal, natural gas or nuclear), generally located outside urban areas, or from renewable sources (hydroelectric dams, wind power, solar parks) and then conveyed to homes and industry through the electricity network (Fig. 7.3).

This is a complex, fragile system on which high demands are placed: it has to perform perfectly 24 h a day and 365 days a year. Few things unsettle us in our daily lives as much as a power cut, and yet, if we come to think of the thousands of kilometers of high voltage cables, the innumerable transformation plants, the dozens of power stations that have to work flawlessly, the thunderstorms, the windstorms that can break the posts or damage the wind turbines, it seems almost like a miracle that power outages are so infrequent. An additional difficulty comes from the fact that *electricity can't be stored, but has to be transmitted and used while it is being generated.* The large power plants can't use accumulators or batteries, unlike motorcars and photovoltaic systems; they can only hold small

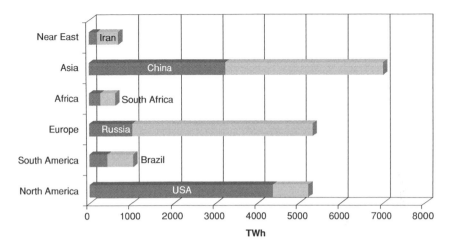

Fig. 7.4 World production of electricity, by regions, with the largest producer. *Source* (BP 2008)

amounts of energy for very short periods. As we have seen, the only practical means to store great amounts of electricity is by using hydraulic pumps. But in most countries—and certainly in Spain—this provides only scant reserves due to limited availability.

Electricity in the World

Figure 7.4 shows electricity generation for 2007, as usually by world regions and expressed in TWh (that is, trillions of kilowatt-hour). Most probably, the reader is already familiar with the landscape. Electricity generation in Europe and in North America is almost identical, around 5,300 trillion kWh in each region. Asia, led by China, stands at almost 6,400 TWh, with an impressive ascending curve reminding of coal consumption (which of course it is tightly related to), while Africa and South America are left far behind. As usual, the USA is number one, generating 22% of all electricity worldwide, a bit less than all of Europe together (about 28% including Russia and the ancient Soviet Union). China generates 16.5% of world electricity.

Considering per capita consumption, again we have the usual pattern. US Americans gobble up 14 million kWh per person, about double as much as the average for Europe, almost six times as much as China, twenty times as much as India and 35 times as much as Africa (without South Africa). It's plain that the Americans don't have a better standard of living than the Spanish, Italian, French or Japanese, nor do they seem to be happier, more long-lived, more cultured, or finer gourmands. However, they consume double as much electricity as in these

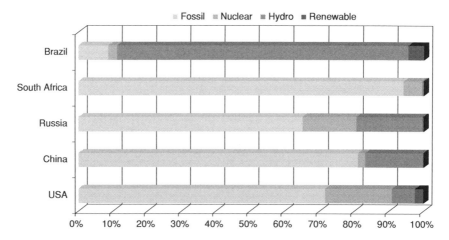

Fig. 7.5 Electricity mix for the countries with the largest electricity consumption

other developed countries. This means the arguments in favor of energy efficiency aren't trivial. A world that made do with 5 MW h per citizen instead of almost 15 MW h would be able to save on coal, and consequently there would be less pollution.

The problem, as we have seen, comes with the second part of the equation. If we want to be honest, we can't deny an equal amount of 5 MW h to the Chinese (that is, doubling the electrical consumption for 1.3 billion people), the Indians (multiplying electricity by seven for another billion) and the Africans (a factor of 10 for 800 million).

Let's look at it from another angle. Squaring accounts in a developed world *that does not waste* (4 MW h per capita, almost the fourth part of the USA, half as much as France, 50% less than Spain) and *is just* (4 MW h per capita for almost 7 billion people) would lead to an electricity consumption of almost 30,000 TW h, 50% more than current figures. Once again, the only valid conclusion is that the world needs more energy than its current consumption if we want to lift the poor from their destitution.

Finally, Fig. 7.5 gives the electricity mix—that is, the share of the different primary energy sources in the electricity generation pie—for the four countries at the forefront of electricity generation. We can see that all of them are highly dependant on fossil fuels, with the exception of Brazil. Among alternative energies, nuclear and hydroelectric have similar shares, between 5% and 20%, while the share of renewables is very small. This raises the key question of this book:

How can we increase the electrical power available—a more equitable world—while reducing consumption of the main resource we use to generate it, that is, of fossil fuels?

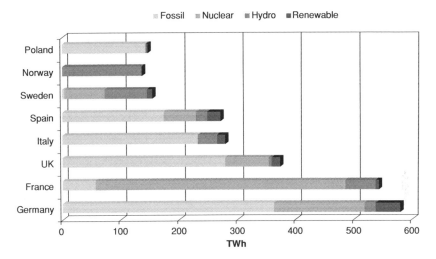

Fig. 7.6 Electricity production in several European countries. *Source* (IEA 2008)

This question is as important as the question *When*? As we have seen, we don't have unlimited time before the shortage of oil and natural gas becomes apparent, and the use of coal worsens the greenhouse effect.

Electricity in Europe

What about the old Continent? How do the developed countries in Europe handle electricity? Figure 7.6 compares some of them. The total length of the bars gives the production in absolute values (in TW h), broken down into four categories: generated from fossil fuels, nuclear, hydroelectric, and renewable.

Each country has its own formula, and there are striking differences. For example, France gets almost 80% of its electricity from nuclear power, which is absent in Italy or Poland (in exchange for an 80% dependency on fossil fuels in Italy and almost 100% in Poland). Norway doesn't use nuclear energy either; here, like in Uruguay or Paraguay, almost all of it is hydro electrical power. The rest of the countries mix different ingredients into the cocktail. Sweden includes 50% nuclear and 50% hydro electrical power, while Germany and the UK go for something similar to the Spanish recipe: 60–70% of fossil fuels, 20–30% nuclear, 10–20% renewables, including hydroelectric.

But we mustn't forget that the United Kingdom can count on plenty of oil and gas from the North Sea, while Poland has important coal deposits. Sweden and Norway boast impressive hydro resources. Germany and Italy haven't much of that, Spain and France even less. Spain is obliged to import practically 100% of the oil and natural gas it consumes, and almost 70% of coal. France is in a similar situation, with even less hydropower than Spain.

In the 1970s, the French government summarized this state of affairs in a famous slogan: "no gas, no coal, no oil: no choice". For 15 years, nuclear energy was developed aggressively, and in 1990 France was already generating 80% of its electricity from the atom, thus spectacularly reducing both its coal imports and its CO_2 emissions.

In contrast, nuclear energy is banned in Italy (though it imports quite a lot from France, probably losing its radioactivity at the border). As a result, the mix is characterized by fossil fuels and the highest energy dependence in the group, similar to Spain. This dependence, particularly when it comes to natural gas from Russia, is a big headache for the country.

References

BP. (2008). BP world statistics. http://www.bp.com/.
International Energy Agency. (IEA 2008). World energy outlook. http://www.worldenergy outlook.org/2008.asp.

Chapter 8
The Bequest of a Supernova

Like bonfires in the night/one every second/the last one
performing its burning dance in the galaxy/triggered Johannes
Kepler's extasis/forever the geometer of mysticism/one more
explosion might have been the end/of the old planet.
And yet/this particular chant/of sound waves and gamma rays/
this blinding and overwhelming way of dying/(dense swan of
carbon, silica and oxygen)/bequeathed to the beginning of
time/a nutritous trace.
Today the scientists/with the joy of ancient poets/they
proclaim/in the days of the totem and the sphere:/we're the
children/of a star

Natalia Carbajosa

Fire Flowers

Hanabi translates as "fireworks", but the Japanese expression is formed by joining
the kanjis "hana", one of whose meanings is "flower", and "bi", "fire". That's
literally "fire flower".

Japanese people, like the Chinese and the Spanish, are great fans of pyro-
technics. I remember one of those festivals, as perfect and minimalist as the
gardens of the town that hosted us, Takayama, a tourist spot in the heart of the
Japanese Alps, about 300 km from Tokyo.

It was the year 1998 and my visit to Takayama was not, strictly speaking, part
of a sightseeing tour. I was taking part in the "Neutrino 98"[1] conference, devoted
to neutrino physics, then and now my professional field.

Among stable elementary particles, neutrinos are—not counting the photon, the
light quantum—the lightest and most elusive. They are a plentiful by-product of
nuclear reactions, in particular of the decay of radioactive nuclei and the fission
processes that will be mentioned in this chapter. They have almost no mass, they
move close to light speed, and matter is, for all relevant purposes, transparent for
them. They resemble ghosts: tiny, flimsy morsels of reality.[2]

And yet, neutrinos play a special role in one of the most violent and crucial
processes in the Universe, the explosion of a supernova.

When they die, some stars, like it will happen to our own sun, expand and turn into
a red giant, and in their death-throes they engulf the planets orbiting next to them.

---

[1] http://www-sk.icrr.u-tokyo.ac.jp/nu98/.
[2] For curious readers: http://www.madrimasd.org/cienciaysociedad/mediateca/default.asp?video
ID=960.

J. J. Gómez Cadenas, *The Nuclear Environmentalist*,
DOI: 10.1007/978-88-470-2478-6_8, © Juan José Gómez Cadenas 2012

Others, more massive and haughty, end their lives with a tremendous bang, such a fabulous explosion that for weeks or months the light emitted by the dying star is stronger than that of its whole galaxy. Every century, one or two supernovas light up in the Milky Way. 1998 was the tenth anniversary of the first—and up to now only— observation of neutrinos coming from one of these supernovas by the Super-Kamiokande detector,[3] a huge apparatus built to study these particles, located in Takayama.

Democritus Revisited

Two and a half thousand years before our conference, the philosopher Democritus proposed one of the most fruitful ideas human thought has ever achieved: he held that everything is composed of tiny, indivisible particles called atoms (from Greek ατομοσ, meaning "uncuttable").

Like it happens to any thinker who is too advanced for his age—and this Greek man was 2,500 years too early—his ideas were ignored by most of his contemporaries, with the only notable exception of Aristotle. It must be said, however, that the Hellene atomists roamed a metaphysical world; in practice, proving their ideas right was no easier than proving the existence of archangels.

Around the year 1661, the "natural philosopher" (as the first scientists called themselves) Robert Boyle suggested that matter is composed of different combinations of "corpuscles" replacing the classical elements (air, fire, earth and water). A hundred years later, in 1789, the ill-fated scientist Antoine Lavoisier[4] refined the concept, defining atoms as matter which cannot be decomposed by chemical means. In 1810, John Dalton went one sep further, proposing that each element is made up of atoms of just one kind.

Another Englishman, J. J. Thomson, discovered the electron in 1897, showing it was a subatomic particle, and thus refuting the idea atoms were indivisible. Thomson pictured atoms like a kind of plum cake consisting of positively charged dough with negatively charged plums (the electrons) scattered all over (or blueberries in a muffin). But 12 years later Ernest Rutherford showed that atoms consist of a positively charged core with negatively charged electrons orbiting around it. The core is 10,000 times smaller than the atom. You can imagine a circular sports stadium where electrons go round in the upper tiers and almost all of the atom's mass in concentrated in a grain of sand in the middle of the arena.

Figure 8.1 represents the modern idea of atom. Inside the atomic nucleus we find *protons* (positively charged particles whose mass is about 2,000 larger than the electron's mass) and neutrons, whose mass is the same as the proton's, but which lack electric charge.

[3] http://www-sk.icrr.u-tokyo.ac.jp/sk/index-e.html.
[4] Whose life was put to an abrupt end by the guillotine of the French Revolution.

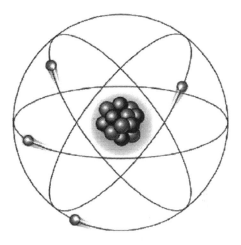

Fig. 8.1 The modern
concept of an atom: the core
contains neutrons and
protons, orbited by electrons

The Atomic Nucleus

Why does the atomic nucleus hold protons and neutrons and not just protons? After all, the chemical properties of an element only depend on the number of electrons the atom possesses (a number we call Z). If we arrange the elements according to their value of Z, we get the famous *Periodic Table* (Fig. 8.2). Z = 1 corresponds to hydrogen, the lightest element of creation, with an electron and a lone proton in the core. Z = 2 means helium, another light, abundant gas. Z = 3 is beryllium, Z = 4 is lithium and so on. Adding just one electron can completely change the chemical properties of an element. For instance, Z = 10 is neon, a noble gas (a gas which doesn't react with other chemicals, that is, which doesn't mingle with plebeians), while sodium, with Z = 11, is a very reactive metal.

If hydrogen (Z = 1) has one proton and one electron (and is therefore neutral, like all ordinary matter, as the electrical charge is balanced), we might expect that helium (Z = 2) has two protons to balance its two electrons, beryllium (Z = 3) three protons and so on. But if we look at it carefully, there's something wrong. Take helium for example, picturing the atom like a stadium: two protons (both with positive charge) huddling in the core, while electrons go round and round in the far away tiers. *But like charges repel each other.* Looking at the stadium from a distant point, the electrons' charge seems to make up for the protons' charge, but if we peek into the nucleus of the atom, the two protons should be pushing each other apart. The situation is worse in the case of lithium and beryllium, not to speak of iron (Z = 26) or lead (Z = 82), where the nuclei are like bunches of mutually repelling grapes. If we think about it, we arrive at the conclusion that the only stable atom is hydrogen (only one proton), and the rest of matter doesn't exist.

Compelled by the fact that this is not the case, physicists had to find some explanation to justify their own existence. The explanation is the so-called *nuclear force* or *strong force*, acting between protons and neutrons, but not between

1 H	2 He

3 Li	4 Be	5 B	6 C												7 N	8 O	9 F	10 Ne
11 Na	12 Mg	13 Al	14 Si												15 P	16 S	17 Cl	18 Ar
19 K	20 Ca	21 Sc	22 Ti	23 V	24 Cr	25 Mn	26 Fe	27 Co	28 Ni	29 Cu	30 Zn	31 Ga	32 Ge	33 As	34 Se	35 Br	36 Kr
37 Rb	38 Sr	39 Y	40 Zr	41 Nb	42 Mo	43 Tc	44 Ru	45 Rh	46 Pd	47 Ag	48 Cd	49 In	50 Sn	51 Sb	52 Te	53 I	54 Xe
55 Cs	56 Ba	57-71 TR	72 Hf	73 Ta	74 W	75 Re	76 Os	77 Ir	78 Pt	79 Au	80 Hg	81 Tl	82 Pb	83 Bi	84 Po	85 At	86 Rn
87 Fr	88 Ra	89 Ac	90 Th	91 Pa	92 U

Lanthinide	57 La	58 Ce	59 Pr	60 Nd	61 Pm	62 Sm	63 Eu	64 Gd	65 Tb	66 Dy	67 Ho	68 Er	69 Tu	70 Yb	71 Lu

Actinide	93 Np	94 Pu	95 Am	96 Cm	97 Bk	98 Cf	99 Es	100 Fm	101 Md	102 No	103 Lw	104

Fig. 8.2 The periodic table of elements

protons (or neutrons) and electrons. The nuclear force doesn't distinguish between protons and neutrons, and it makes them cling to each other with a strength that is a hundred times the strength of electric repulsion. But in exchange, this force disappears once you leave the tiny space occupied by the atomic nucleus.

To sum up: the nuclear force is the mortar that binds protons in the nucleus, in spite of the electric repulsion that tries to tear them apart.

What about the neutrons? They're like some kind of extra plaster added to the nucleus to keep it stable. Neutrons lack an electric charge, so they don't repel each other and don't repel protons either. In a way, they muffle the protons against the electric repulsion and thus contribute to the atom's stability.

In the case of helium, there are two extra neutrons accompanying the protons, which intensifies the action of the strong force against the electric repulsion. Similarly, lithium ($Z = 3$), with three protons in the core, should hold three neutrons and so on. Or should it?

Isotopes

Not quite. In fact, light elements (hydrogen, helium, beryllium, boron, carbon) have an (almost) identical number of protons and neutrons. But as we move towards heavier nuclei, we need more and more plaster to make up for the enormous repulsive force exerted by the clusters of 26 (iron), 82 (lead) or 92 (uranium) protons. In the case of uranium, we need not fewer than *one hundred and forty something* neutrons to "stabilize" the 92 protons.

Just a moment. Why *one hundred and forty something*? That must be a specific number, like 142 or 143 or 146, mustn't it?

Well, when it comes to adding neutrons, nature doesn't use a rigid formula. For helium, two protons and two neutrons are enough, but in the case of lithium there are already variants: there's one with 3 protons and 3 neutrons (we call it lithium-6, Li-6 or ^{6}Li, where 6 stands for the global figure of protons and neutrons) and another with 3 protons and 4 neutrons lithium-7, Li-7 or ^{7}Li). These two versions of lithium are identical from the point of view of chemistry (as chemical properties only depend on the number of electrons, $Z = 3$, in this case), but their physical properties differ. In nature, lithium is made up of 92.5% ^{7}Li and 7.5% ^{6}Li. That's what we mean when we say that both are *isotopes* of lithium.

Natural uranium comes as a mixture of three isotopes. The most abundant is ^{238}U, with 92 protons and 146 neutrons. About 0.7% of it is ^{235}U, with 143 neutrons, and a very small fraction is ^{234}U, with 142 neutrons. The key to nuclear energy is the fact that the isotope ^{235}U may be split into two smaller pieces when hit by a slow neutron, and energy is released in the process.

Stardust

But let's leave nuclear energy aside for the moment and let's return to what happens inside the stars; the reaction that takes place there releases much more energy than splitting ^{235}U. Here we're dealing with *nuclear fusion*, where two protons and two neutrons join to form helium. In this process, a million times more energy is produced than in a conventional chemical reaction. However, for these fusions to happen, first we have to overcome the electric repulsion between the two protons to be joined, and this requires providing them with a lot of energy. In an H-bomb, the explosion of a conventional atomic bomb supplies this initial energy. In a star, it comes from the high temperature of the plasma that makes up the star.

Helium, being heavier than hydrogen, tends to accumulate in the deeper layers of the star's atmosphere. If we picture the star like an astral onion, hydrogen, on the other hand, is to be found in the outer skin. Below it there's helium, which is in turn subjected to fusion reactions creating lithium, and then beryllium, carbon and so on: increasingly heavier elements that build up in increasingly deeper layers, right down to iron, which accumulates in the star's core.

Each fusion reaction releases energy due to the fact that the resulting nucleus is more stable than the parent nuclei which are joined. Iron is the most stable element in nature, so that's the point where the machinery operating in the giant stellar cauldron comes to an end.

There comes a moment when the immense energy released by the fusion reactions is unable to hold the colossal mass of the star and it collapses. A myriad of neutrons are created in the process, and their reactions at the heart of the supernova gradually form the elements heavier than iron. A neutrino "shockwave" is also produced. Paradoxically, in the infernal core of the boiling star, these almost inert particles manage to destroy the outer layers of the star. Like a bursting grenade, the supernova, in its death rattle, spews out tons of matter including practically all chemical elements. The neutrinos escape, taking with them the memory of the dying fire flower.

Millions of years later, the leftovers of the supernova, attracted by relentless gravity, start to condense and sometimes form a new star, which may come along with a solar system. The stardust solidifies and planets are born, literally made up of the matter created inside the supernova. The Earth is one of those planets, and all of the metals our civilization uses so lavishly were brought forth by a mother sun that existed before the sun. Among them is uranium, thorium and the other radioactive elements whose decay keeps our planet's core cooking.

The Discovery of Radioactivity

In 1896, while studying the fluorescence of uranium salts, Henry Becquerel accidentally discovered radioactivity. He had placed uranium between two photographic plates while setting up the experiment, but before he had carried it out he realized that these plates—which he had wrapped in black paper to protect them from light—had darkened as if they had been exposed to the sun. He was forced to conclude that the uranium salts were spontaneously emitting radiation.

Marie and Pierre

Thirty years before, in 1867, Maria Skłodowska was born in Warsaw, the fifth and youngest child of well-known teachers Bronisława and Władysław Skłodowski. In 1891, 24 years old and broken hearted,[5] Marie arrived in Paris and enrolled at the Sorbonne, attending classes in the morning and tutoring in the evening to earn her keep. Two years later she was awarded a degree in physics and started work in an industrial laboratory at Lippman's. But at the same time she continued studying and earned a degree in mathematics in 1894.

Marie, Marie. Decades later, her would-be lover, then rector of Kraków University, would still remember her and spend hours contemplating her statue. She was a poor Cinderella, with no other fairy godmother than her talent and courage. There were no women's rights associations, no ministry of women's affairs in 19th century France, where the young girl—almost an old maid by the standards of the time—now worked on her PhD. Female, poor, a foreigner. In the portraits she dresses in black, wears her hair tied up, her look is intense, almost tragic. A fragile woman, all mourning, bones and courage.

---

[5] Because of an impossible love with a man who would become the eminent mathematician Kazimierz Zorawski. Marie worked as a householder at the Zorawskis, cousins of her father, and the young man's familiy opposed the engagement with a poor relative.

In 1896 she meets Pierre Curie, a physicist like herself, but she doesn't build up her hopes. She wants to return to Warsaw, work in her homeland, and she doesn't count on him joining her. Fortunately, Krakow University lends her a hand, turning down her application—a most reasonable decision based on the fact that she is a woman. She returns to Paris and marries Pierre. Cinderella has met her prince, who, instead of giving her a pair of glass shoes, offers her a scientific instrument of his own invention, the Curie electrometer. This was a device measuring very low electric currents, and with it Marie discovered that uranium salts electrify the air surrounding them. From this she was able to establish that the *activity* of these salts depended exclusively on the amount of uranium present, so the radiation didn't stem from some interaction of molecules, but from the atom itself.

During the following years Marie studied pitchblende (a mineral mostly made up of uranium oxide, UO_2). Her electrometer proved that this mineral was four times as active as uranium itself. She concluded that pitchblende must contain some other substance far more active than uranium. According to her biographer, Robert (1974)

> The idea was her own; no one helped her formulate it, and although she took it to her husband for his opinion she clearly established her ownership of it. It [is] likely that already at this early stage of her career [she] realized that... many scientists would find it difficult to believe that a woman could be capable of the original work in which she was involved.

How did Prince Charming react to the crazy ideas of his wayward Cinderella? If this story weren't a fairy tale, he might have felt jealous, or even worse, he might not have taken her ideas seriously. But instead of that, Pierre put his work on crystallography aside in order to help her. In April 1898 they began their experiments, processing 100 g of pitchblende in search of the new element postulated by Marie. How naïve they were! What they were searching for was present in such minute quantities that they had to process *tons* of ore before they were finally able—in July 1898—to separate polonium (thus named in honor of her native country) and some months later—in December 1898—radium.

It took them more than a ton of pitchblende to obtain a tenth of a gram of radium chloride in 1902, and not until 1910 did Marie, then without Pierre, separate the pure radium metal. In spite of her arduous efforts and of having come across a totally new technique, Marie Curie refrained from patenting the radium-isolation process, so that the scientific community could do research unhindered.

Marie without Pierre. In 1906, the scientist, husband and inseparable partner was accidentally killed by a horse drawn vehicle. In 1903 both of them had received the Nobel Prize in Physics, together with Henri Becquerel. Maybe it wasn't a fairy tale after all, or perhaps you have to finish the tale in time if you want a happy end.[6]

---

[6] But perhaps Stephen Mitchell, Rainer Marie Rilke's great translator, was right when stating that after God's visit a happy ending is unnecessary.

These paragraphs can't but betray the great admiration I have felt all my life for this extraordinary lady. It would be decades before people started talking about women's liberation, and she managed to win two Nobel Prizes (the Nobel Prize in Chemistry in 1911) and was the first woman to be a lecturer in the Sorbonne. Her daughter Irene and her son-in-law Fréderic also won the precious award.

What follows is for those readers who don't like too rosy tales. In 1911 it was revealed that 5 years after her husband's death Marie had had an affair with the physicist Paul Langevin, who later abandoned his wife. In spite of her scientific stature and all her personal and scientific generosity, the press attacked her. She was accused of being a man-eater—she was 5 years older than Paul—of destroying happy families, of being a Jew—the Dreyfus affair was still in the air. Time passed. Marie Curie continued with her life, her career and her contribution to a better world, among them the development of mobile radiography units for the treatment of wounded soldiers in World War I. She received a heap of prizes and still, to quote Albert Einstein, she was "the only person I know who has not been corrupted by her fame". She died in 1934, with 67 years, a premature death probably related to the radiation she and Pierre had been exposed to during their research. Her work was groundbreaking.

Alpha, Beta and Gamma Particles

Radioactive nuclei like radium and polonium decay (that is, they are transformed into lighter nuclei) by emitting various kinds of radiation that we refer to by their historical names, the first three letters of the Greek alphabet: alpha radiation, that is, two helium nuclei (two protons and two neutrons); beta radiation (electrons) and gamma radiation (light quanta, or high energy photons). When it comes to penetration depth, alpha particles are the easiest to stop: it's enough with a sheet of paper, a few centimeters of air, the human skin. Beta particles, or electrons, have a higher penetrating power, but can be shielded against with a few millimeters of aluminum. They can burn the skin, but don't penetrate human tissue. Consequently, neither poses a health risk, except when ingesting a highly radioactive product, in which case they can cause inner injuries.

Finally, gamma radiation is similar to X-rays, but can be much more energetic. In order to stop it, you need several centimeters of lead or iron, so the human body can be wounded if it is directly exposed to a very concentrated source (Fig. 8.3).

Radioactive Decay

As we have seen, there's a competition between the (attractive) nuclear force, which contributes to the stability of the atomic nucleus, and the electromagnetic force (repulsive between protons), which tries to tear it apart. In light nuclei, the

Fig. 8.3 Penetrating power
of alpha, beta and gamma
radiation

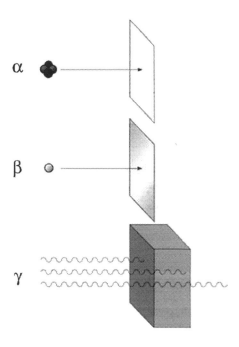

strong force has the upper hand, but as the number of protons increases, we have to add extra "plaster" (that is, more and more neutrons) to keep the nucleus from breaking up.

However, there comes a moment when the neutrons aren't up to it. They just can't keep the nucleus stable in the long run, so it transmutes into another, lighter element, emitting alpha, beta or gamma particles in the process. These elements are called *radioactive*.[7] All elements with a nucleus heavier than bismuth (83 protons, 126 neutrons) are radioactive.

What is meant by "activity" is the number of fissions (decay), and thus the number of particles emitted by a sample of material.

We often read, generally in the context of anti nuclear propaganda, about the grave risks related to "highly radioactive" waste that "lasts for thousands (or millions) of years". This is the origin of one of the most widespread fears, the idea that radioactive waste is capable of scorching us, not only right after being taken out of the nuclear reactor, but 10,000 years later.

The fact is that, by definition, if an element is radioactive, it ceases being dangerous after a relatively short time span, for the simple reason that it disappears, being transmuted into another element. In contrast, if a sample of the material is still active after 10,000 years, this means the element it is made of is only lightly radioactive.

---

[7] The root *radio* obviously refers to the element discovered by the Curies.

I would like to present a parable to make this point more comprehensible.

Two citizens, H. Scrooge and C. Spendthrift, receive bequests from their rich uncles. Each of them inherits a fixed amount of one million Euros, and the will specifies that they will be paid a monthly stipend *which is a fixed percentage of the global sum available each month*, not a fixed salary. Together with the will, the executor hands them a form where they have to put down the value of this percentage.

Mr. Scrooge gives it a long thought and goes for 1%. Spendthrift, however, who's already planning to buy an apartment and a new car, claims 5%. Let's see how much money each of them receives.

In the first month, Scrooge is paid 1% of 1,000,000 Euros, that's 10,000 Euros. The amount remaining after his stipend has been drawn is 1,000,000 − 10,000 = 990,000 Euros. In the second month, he gets 1% *of the remaining capital*, that is, 9,900 Euros, which is deduced from the money left and so on. The stipend drawn by Scrooge is not very large, so the initial capital decreases slowly, and so do the monthly stipends. After six months, he still gets 9,509 Euros; after a year, 8,953; after two years, 7,936; after four years, 6,235 etc.

On the other hand, Spendthrift receives 5% of one million, not less than 50,000 Euros, with which he purchases a convertible car. Six months later he gets 3,689 Euros, much more money than the measly 9,509 Scrooge is paid, and the happy Spendthrift books a holiday in the Caribbean. One year later, he still gets a monthly allowance of 28,440 (which he uses to decorate his new apartment), and two years later, 15,367, double as much as Scrooge, though that's not enough to pay the installments of his yacht. Four years later, he receives only 4,487, less than Scrooge, and the creditors are after him.

Figure 8.4 (above) shows our characters' finances. In the first month, Spendthrift honors his name and spends much more than Scrooge. However, as time goes by, his lifestyle takes a toll on him, and after four years there's so little money on his account that his monthly stipend is slightly below Scrooge's. After eight years, Scrooge still receives more than half his initial stipend, while Spendthrift is broke.

Figure 8.4 tells the same story from another point of view. The difference between Spendthrift and Scrooge lies in the speed at which the former's legacy is reduced, being almost gone after four years, while the latter still keeps half of the capital at this moment.

Radioactive nuclei behave in exactly the same way as our fictitious inheritors. The number of atoms that decay per unit of time (seconds, months or years) is a constant fraction of the number of atoms present in the sample at each moment, and this fraction depends on the radioactive material.

An element which is only lightly radioactive behaves like a Scrooge. If the initial capital is the initial number of atoms in the sample and the percentage defining the monthly stipend is the decay rate, then a stingy radioactive element loses its nuclei gradually and in exchange emits a small number of alpha, beta or gamma particles (compared with the number of nuclei contained in the sample), which slowly becomes larger over time. In exchange for its restraint, a lightly radioactive element keeps emitting particles after a very long time.

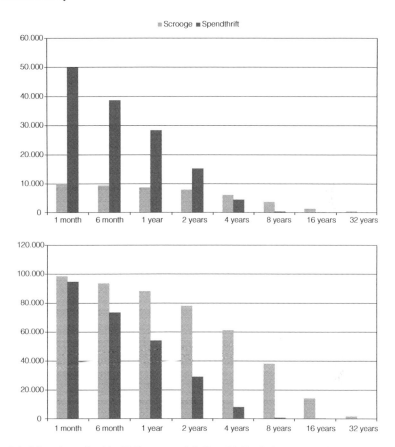

Fig. 8.4 Stipends received by H. Scrooge and C. Spendthrift; (below) Remaining capital for both

In contrast, a very radioactive element draws a high monthly stipend at the beginning (that is, it emits a lot of alpha, beta or gamma radiation), but it runs out of capital (that is, of nuclei) very quickly, so after a shorter time span it stops emitting. The more radioactive the element is, the sooner it ceases to be so.

If we study Fig. 8.4, we can see Scrooge's capital is reduced to half after about 9 years, which is much longer than the time Spendthrift needs to squander half his fortune. That would be the equivalent of an element's *half-life*: the name physicists give to the period of time necessary for a radioactive sample to be reduced to half. In our example, the difference between a monthly rate of 1% (Scrooge) and 5% (Spendthrift) corresponds to a half-life of 9 years for the former and one year for the latter.

Now, when it comes to radioactive elements, ^{238}U is Moliere's miser personified. Its half-life is not less than 4,500 million years, a third of the age of the Universe, more or less the age of the Earth. This means that on the bank account we've still got half the ^{238}U the supernova deposited to it when it created the planet. In contrast, elements like caesium-137 (^{137}Cs) or strontium-90 (^{90}Sr) have half-lives around 30 years, which makes them real wasters. As we'll see in the next chapters, both accumulate in used fuel, and during the first years their short half life (high decay rate) gives rise to a very strong gamma particle emission. However, after 100 years the activity (the monthly stipend, that is, the number of particles that are emitted) has been reduced by a factor of 100, and after 300 years, by a factor of 1,000.

A radioactive source can be characterized by the number of particles it emits per second. The unit of radioactivity is the Becquerel, Bq (named after Henry Becquerel), which amounts to one decay per second. The activity of one gram of ^{238}U is about 12,000 Bq, while the activity of one gram of ^{235}U is considerably higher (given its much shorter half life of 700 million years): 79,000 Bq. But that's next to nothing compared to the activity of 1 g of ^{137}Cs during the first year after coming into being: almost 2 trillion Bq (1.8 Tbq). In terms of uranium, we would need 1,500 tons to generate this amount of particles.[8]

An Unexpected Surprise

The discovery of radioactivity, as the Curie couple were well aware from the beginning, amounted to a revolution in the scientific understanding of the properties of matter. Some years later, in Rome, Enrico Fermi, one of the greatest physicists of the 20th century, took to bombarding different elements with neutrons, and observed that in almost all cases the result was a radioactive nucleus that decays by emitting beta particles. Fermi was, in fact, duplicating the process that takes place during a supernova explosion, creating elements of increasing weight by means of neutron capture. Eventually this led him to uranium and to even heavier elements (the so-called trans-uranians, which we will frequently encounter in the following chapters).

Soon other outstanding physicists were reproducing Fermi's experiments, among them the German chemist Otto Hahn and his colleague, the physicist Lise Meitner. At the same time, in Paris, the couple formed by Frederic and Irene Joliot-Curie followed in the steps of the previous generation, with sophisticated experiments aimed at analyzing the composition of substances irradiated by neutrons. In 1938 they published a peculiar finding. They had found lanthanum among the products formed by irradiated uranium.

What was peculiar about that? The atomic mass of lanthanum (58 protons) is almost half of the uranium mass, too light to be a product of uranium. As we have seen, uranium is lightly radioactive and decays by emitting an alpha particle and

---

[8] Here we're ignoring the fact that ^{238}U decays into other radioactive elements, some of them with quite short half lives, each of the contributing to the total activity of the sample.

changing into thorium, which in turn decays into other lighter elements, including radium, bismuth and polonium, up to the stable lead isotope, lead-206, which is absolutely stable and whose mass is much larger than lanthanum. The presence of lanthanum among the products of irradiated radium was as inexplicable as gold nuggets appearing in a cement block.

And that's not all. Otto Hahn and his colleague Friedrich Strassmann managed to prove, in 1939, the existence of an element that was still lighter than lanthanum: barium. Otto had an intuition of a possible explanation, but he wasn't confident, so he wrote to his friend Lise.

Lise and Otto

Lise had been born in 1878, the daughter of a Viennese lawyer of Jewish background. In those days, Austrian women were banned from institutions of higher education, but the young lady was fortunate to count on the support of a progressive, influential and wealthy family. In 1901 she enrolled at university to study physics. At the turn of the century, Vienna was one of the cultural, scientific, artistic capitals of the world: the city of Klimt, Gödel, Wittgenstein, Freud and many others. The physics chair was held by the eminent Ludwig Boltzmann, and the department was boiling with activity thanks to the new radioactivity science. It wasn't exactly the place where you would expect to find a shy Jewish girl. Except for the fact that she was the best student in her class, who managed to obtain her doctoral degree in 1905, the second woman to accomplish this feat in her country.

In the enlightened Vienna ministries of gender and equality weren't yet in place, and the best offer Lise received when finishing her doctoral thesis was to work in a gas lamp factory. She rejected it, and backed by her father she went to Berlin, the realm of another giant scientist, Max Planck. The great man was not exactly a feminist; up to then, he had allowed no women to attend his lectures. But he made an exception with the young Viennese lady, who soon after became his assistant.

In Berlin, Lise met Otto Hahn and became his friend and professional colleague, a relationship that in some ways echoes the one that joined Pierre and Marie. We will see, however, that this one had a bitter end.

In 1912, after having worked for several years and published plenty of scientific articles with Lise, Hahn was appointed professor at the newly founded Kaiser Wilhelm Institute for Chemistry in Berlin. Lise received a position as "guest", without a salary, and this situation continued until 1913, when she was already 35 years old and had acquired an excellent scientific reputation.

In 1917, Lise and Otto discovered the first long-lived isotope of the element protactinium. In 1913 she discovered the cause of an important effect called Auger electron emission. The name is in honor of Pierre Auger, a French physicist who independently discovered the effect *two years later*. This would not be the only time that the scientific establishment denied Lise the credit she deserved.

In 1933, when Hitler came to power, Lise held a managing position in the Institute. Her Austrian citizenship protected her from Nazi inquisition, while all other Jewish scientists, including her nephew Otto Frisch and other eminent figures such as Fritz Haber and Leó Szilárd were dismissed, most of them emigrating from Germany. Lise decided to ignore the situation and bury herself in her work, an attitude she would regret for the rest of her life.

Five years later, she was almost caught in the Nazi trap. In July 1938 she escaped to the Netherlands with the help of the Dutch physicists Coster and Fokker and, in her own words, with ten marks in her purse. She was also wearing a diamond ring her friend Otto Hahn had inherited from his mother and had given her in case it was necessary to bribe the frontier guards. Lise escaped by a hair's breadth, as Kurt Hess, a colleague, a chemist who was an avid Nazi, had informed the Gestapo about her plans.

Lise took up a post in Stockholm and started work with Nils Bohr, who in turn still corresponded with Hahn and other German scientists. In fact, Lise and Otto met clandestinely in Copenhagen in November and surrendered to their particular forbidden passion: jointly planning a new round of experiments. They subsequently exchanged some letters. As we have seen, Otto had been able to separate barium and had speculated on a possible cause that he didn't quite dare to articulate. He wrote to his friend and colleague: "maybe you can suggest some fantastic explanation".

Lise Meitner and her nephew Otto Frisch in fact found the fantastic explanation Hahn was asking for. This was nothing less than *nuclear fission*, a process that just a few years later would lead to the making of the atomic bomb, but would also usher in the use of nuclear energy for peaceful purposes. When the minority uranium isotope, ^{235}U, is irradiated with neutrons it breaks up into lighter fragments, among them barium. During the process, other neutrons are emitted that can in turn break up other 2^{35}U nuclei, and an enormous amount of energy is released, according to Einstein's formula, $E = mc^2$.

But this one is not a fairy tale. Otto Hahn wasn't as generous as Pierre Curie and he took all the credit for the discovery of fission, though it was Meitner and Frisch who hit upon the explanation for the experimental results.

In 1944 Otto Hahn received the Nobel Prize for Chemistry for his discovery of nuclear fission. While Marie Curie's merits were recognized by two Nobel Prizes, Lise's weren't sufficient for the Stockholm committee to reward her together with her old friend. It's true that later in her life she was granted many prizes that in some way made up for the injustice, among them the Enrico Fermi Award in 1966. Perhaps the most important recognition came in 1997, when element number 109 was named meitnerium in her honor.

Her friendship with Otto had an unhappy ending. After the war, Lise criticized Hahn and other German scientists bitterly, accusing them of having collaborated with the Nazis and not having contested the crimes of Hitler's regime. In a letter to Hahn she wrote:

You all worked for Nazi Germany. And you tried to offer only a passive resistance. Certainly, to buy off your conscience you helped here and there a persecuted person, but millions of innocent human beings were allowed to be murdered without any kind of protest being uttered... first you betrayed your friends, then your children in that you let them stake their lives on a criminal war—and finally you betrayed Germany itself, because when the war was already quite hopeless, you did not once arm yourselves against the senseless destruction of Germany.

Lise became a Swedish citizen in 1949, but she lived in the UK from 1960 on and died in Cambridge in 1968. Her nephew Otto Robert Frisch composed the inscription on her headstone. It reads "Lise Meitner: a physicist who never lost her humanity."

Five Minutes to Midnight

The fission of ^{235}U is represented in Fig. 8.5. When this nucleus absorbs a neutron, it turns into an unstable isotope (^{236}U), which in turn immediately decays into two smaller fragments, for example krypton (^{92}Kr) and barium (^{141}Ba). Further neutrons are released in the process, and these may in turn strike other ^{235}U atoms and cause a chain reaction. Each fission releases a small amount of energy, about 3.2×10^{-11} J. But the number of atoms contained in a small amount of matter is very large: a kilogram of 235U contains 2.56×10^{24} atoms.

So the splitting of this kilogram releases

$$3.2 \times 10^{-11} \times 2.56 \times 10^{24} = 82 \times 1012 \text{ J} = 82 \text{ TJ} \sim 20 \text{ kt}$$

This amounts to the energy we obtain by burning about 3,000 tons of coal.

There's another, more sinister way to express this colossal energy. The kiloton (kt) stands for the energy equivalent to the explosion of one thousand kilogram of dynamite (or TNT, to be more precise). The energy released by the fission of 1 kg of uranium is equivalent to twenty thousand tons of dynamite, or 20 kt.

In their 1939 article, Lise Meitner and Otto Frisch had already taken note of the tremendous amount of energy released by the fission of uranium. French, British, North American, Russian and German physicists immediately understood the implications of a physical process that was able to instantaneously release as much energy as thousands of tons of dynamite. In the USA Leó Szilard, Edward Teller and Eugene Wigner persuaded Albert Einstein, who was already a celebrity at the time, to write the famous letter to the President of the United States, Franklin D. Roosevelt, that led to the Manhattan Project. Lise refused to take part. In her own words: "I will have nothing to do with a bomb!".

Others didn't share her view. It's difficult not to agree with them. Germany had some of the best nuclear physicists in the world, and access to the uranium mines in occupied Czechoslovakia. In North America it was feared that the Nazis might get the bomb ready before them, and this sparked one of the most spectacular intellectual and technological efforts in history. Among those who took part in the

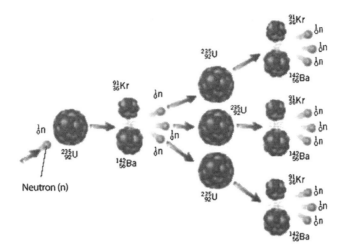

Fig. 8.5 Chain reaction started by the fission of ^{235}U

Manhattan Project there were eight Nobel Prize winners, and 12 of the scientists involved would be awarded the prize in the following decades. In just 28 months, fission moved from the realm of ideas to the terrible reality. On 16th June 1945 the first atomic bomb exploded in Alamogordo, New Mexico. The explosion, which also ushered in the era of open air nuclear tests, released 21 kt of energy, a bit less than the bomb that destroyed Hiroshima on 6th August 1945 and as much as the one that extinguished Nagasaki on 9th August.

In the second volume of his beautiful trilogy "His Dark Materials", Philip Pullman tells us about a magical knife, able to make its way through parallel universes, cutting into the fabric of reality. It's a tool that is needed to save the world, but at the same time it's dangerous by its own nature. Without considering the use humans want to put it to, the weapon intends, above all, to hurt.

Similarly, the bombs that were developed to prevent the tyranny of a nuclear Hitler ended up being used for a very different purpose. They may have brought the war to a quick end; they even may have saved more lives than they destroyed. But it was also said that they had to explode *because that's what they had been built for*. It was in their nature.

Or maybe it was in our nature to lean into the abyss, without knowing whether we would finally drop into the chasm. The Hiroshima and Nagasaki bombs initiated one of the weirdest episodes in history, aptly named MAD (Mutual Assured Destruction). On 29th August 1949 the Soviets were trying out their first atomic bomb, whose design was due to espionage—Klaus Fuchs, a British physicist who had taken part in the Manhattan Project, had passed on the information (Smil 2006).The first fusion bomb (or H bomb, H standing for hydrogen, whose fusion produces helium and generates the explosion) went off in 1952, and the energy released was 10.4 megatons (Mt), about two thousand times the Hiroshima bomb.

In 1961, the USSR responded to this exhibition of destructive power with the most powerful explosion to date, equivalent to 50 million tons of dynamite. The explosion could be seen at a distance of thousands of kilometers, and it released twenty times more energy than the total of all the bombs thrown off during World War 2. Fifteen months later, the Soviet leader, Nikita Khrushchev, revealed that his country possessed a still more powerful bomb, capable of releasing 100 Mt.

Between 1959 and 1961 the USA built 19,500 nuclear warheads, that's a pace of 75 per day. Altogether, they made 70,000 bombs between 1945 and 1990, with Russia settling with the modest figure of 55,000. More than a third of those bombs had a destructive power of between 100 and 550 kt, and were aimed at important cities and relevant military installations. The numbers of casualties that were mentioned in military circles at the time were in the tenths of millions. The Bulletin of the Atomic Scientists, among whose contributors were physicists from all over the world, worried by the infinite foolishness of the arms race, created a logo which perfectly defines those decades: the Doomsday Clock, with the hour hand set at twelve, and the minute hand measuring the time left for midnight, the end of civilization. In 1953, with the H-bomb nuclear tests in the USA and the USSR, the hand was set at two minutes to doom. In 1991, at the end of the Cold War, the editors set the hand back to 17 min. Climate change, terrorism and environmental worries are to blame for the fact that in 2009 we are again at only 5 min to midnight.

The possibility that nuclear countries—such as India and Pakistan—start an atomic war, the phantom of an eventual Israeli attack on Iran—to prevent this country form developing its own bomb—or the fear that terrorists might somehow get their hands on a nuclear warhead seems to justify the little time left for disaster by this clock.

A Nuclear Reactor is not an Atomic Bomb

Consciously or unconsciously, the persuasion that the core of a nuclear reactor is potentially an atomic bomb, ready to burst in case of an operating mistake or a gang of audacious terrorists forcing the operators at gunpoint to take out the control rods, is still embedded in many people's imagination. In fact, though the bomb and the reactor are based on the same physical principle, the way in which they release energy is totally different.

The starting point is the same: a chain reaction. A neutron splits a ^{235}U atom, releasing two or three neutrons, which in turn can split other uranium atoms and amplify the process (Fig. 8.5).

But we mustn't forget that only ^{235}U is split. The majority isotope of uranium, ^{238}U, not only isn't fissile, but on the contrary, it tends to absorb the fast neutrons emitted by ^{235}U when splitting. In natural uranium, the ratio of ^{235}U is very small and the chain reaction can't be sustained given that the much more abundant ^{238}U absorbs the neutrons before they can trigger new fissions.

An atomic bomb is nothing but a chunk of almost pure ^{235}U. To obtain it, you have to separate the ^{235}U and ^{238}U that make up almost all of natural uranium, and gather ^{235}U until you have enough of it. When there's no ^{238}U, the fast neutrons aren't absorbed and so they end up splitting ^{235}U atoms.

And yet, the neutrons can still escape from the uranium without hitting on a 2^{35}U atom if the volume (and hence the mass) of the object is not large enough. In other words, you need a certain *critical mass*; below this value, the reaction isn't sustained, not even with pure ^{235}U.

A conventional atomic bomb is made from two fragments of ^{235}U or plutonium which individually don't reach critical mass, but together exceed it. In order to make it burst, you can use chemical explosives that compress the two fragments, and thus an uncontrollable chain reaction starts immediately.

In a nuclear power plant, there's only 3% of ^{235}U. The chain reaction can never be triggered the way it happens in an atomic bomb because 97% of the uranium is still made up of ^{238}U, and in fact it must be enriched, as we will see in Chap. 9. Besides, the fuel is spread over a very large volume, so it's impossible that it might reach the critical mass necessary for a nuclear blast. An accident may happen if the temperature rises up to the point of melting the core, and in this case there may be a (chemical) explosion due to the release of overheated gas. Accidents in reactors will also be dealt with in Chap. 9.

Plutonium

When ^{238}U absorbs a neutron, it turns into a new element, plutonium-239 (^{239}Pu), which is also radioactive and has a very long half-life of about 24,000 years. In a nuclear reactor, ^{238}U capturing neutrons gives rise to the formation of ^{239}Pu, which slowly accumulates in the fuel rods.

^{239}Pu is a special element—and some would say it's especially evil—for three reasons. First, it doesn't exist in nature, because its half-life, which may seem long by human standards, is ridiculous in terms of geological time. The only plutonium there is in the world has been artificially created by humans. Secondly, it can be split by neutrons, the same as ^{235}U, except that it doesn't require slow neutrons and is therefore more efficient than ^{235}U when it comes to sustaining a chain reaction. That's why it's an extraordinary fuel and not less extraordinary raw material to make nuclear weapons. Finally, we can't avoid producing ^{239}Pu and other isotopes heavier than uranium (called trans-uranians) in a conventional nuclear reactor. Trans-uranians have very long half-lives (by human standards) and are the reason why radioactive waste has to be stored for thousands of years in geological deposits. All this has turned plutonium and its friends into *personae non gratae* in many social circles. And yet, as we will se, we can redeem these pariahs, thanks to an invention that reproduces exactly the ambitions of ancient alchemists: the fast neutron reactors, the genuine philosopher's stone of nuclear energy (Chap. 9).

Peaceful Uses of Radioactivity and Nuclear Energy

It's undeniable that a knife is made for hurting. But it's no less true that a scalpel is made for curing. A piece of iron can be turned into a sword or a plough. There's no doom, no malediction that obliges us to renounce the peaceful uses of nuclear energy in order to avoid past mistakes or get rid of our sins. Some of these uses are invisible but very important, such as the use of radioactive isotope ^{241}Am in smoke detectors installed in public buildings and many private homes. Others have saved millions of lives, starting with X rays (identical in their nature to the dreaded gamma rays) and continuing with the science that in my student days was called "nuclear medicine", a name rarely used in our politically correct times. The famous and still futuristic "hydrogen economy" may find in nuclear energy a basic pillar, and the spaceships of the future will be nuclear-powered.[9]

Nuclear Medicine

This expression, which has already become old-fashioned, encompasses two kinds of activities: the diagnosis of diseases and their treatment. In Europe alone there are about 10 million interventions per year that can be considered nuclear medicine, and radio pharmacy is a flourishing industry in our days.

Diagnosis

Diagnosis techniques are based on an apparent blasphemy. The patient swallows, or is injected, a radioactive substance emitting gamma rays. These gamma rays are emitted from inside his body; they pierce his tissues and reach a camera that records them. The camera uses this information in order to create, with the help of sophisticated software, a 3D image of the region or organs to be studied.

How is it possible? What about the *destructive power of radioactivity*, one of the most repeated clichés of the converts to anti nuclear religion? Our great-grand-mothers already knew the answer: the dose makes the poison, not the substance. A little glass of wine is good for your health. A bottle of whisky for breakfast, not quite so good. A few drops of laudanum help us sleep. An entire vial dispatches us. The doses (Chap. 10) of radioactive substances these patients take in are minute in view of their effects on health, but they are high enough to allow diagnosis.

The most sophisticated process nowadays is PET, Positron Emission Tomography. For this kind of diagnosis, the patient is injected a positron emitting radionuclide. The positron is an antimatter particle, the image of an electron in a

---

[9] Or almost of the present: for example http://www.telegraph.co.uk/scienceandtechnology/science/sciencenews/3322276/Star-Trek-style-ion-engine-tofuel-Mercury-craft.html.

weird cosmic mirror. Like fatal lovers, the encounter of an electron and a positron leads to their mutual annihilation, while two gamma rays are emitted that escape in opposite directions and are detected by the apparatus surrounding the patient. The most common application of PET is non-invasive cancer diagnosis, but it is also used to obtain images of the heart and the brain.

The typical dose in a nuclear medicine treatment such as PET is around 4.6 mSV (see Chap. 10), about double as much as the annual average due to natural causes.

Therapy

A cancer is just a group of cells dividing very fast. The technique for controlling, and often curing, cancers consists in fully harnessing the power of radiation—yes, now it can be beneficial.

This radiation can be external (using cobalt-60 sources or accelerators that produce a very collimated beam of gamma rays focused on the cancerous zone) or internal, by implanting a gamma or beta emitter in the zone to be acted on. One of the most commonly used ions for the treatment of thyroid cancer is iodine-131, while iridium-192 is employed for head and breast cancer. So it's about focusing the poison to cure the patient. In the case of leukemia, the cure is still more drastic; the treatment often requires a bone marrow transplant. Previously, the patient is administered a lethal radiation dose, to annihilate the cancerous marrow. Obviously, the patient doesn't die thanks to the transplant.

What are Radioisotopes, and How are They Produced?

To sum up, nuclear medicine uses a wide range of *radioisotopes* both for diagnosis and treatment. The name is given to radioactive isotopes of a certain element (same number of protons, different number of neutrons). Medical treatment, and other applications, nowadays use more than 200 radioisotopes, most of which are obtained artificially. The most common technique to produce radioisotopes is bombarding stable nuclei (iodine, strontium, cesium etc.) with neutrons inside a nuclear reactor.

These reactors aren't the huge commercial installation for the purpose of generating electricity, but specially built devices called research reactors, much smaller and easier to operate than their giant cousins.

The Dose Makes the Poison

^{241}Am is a very radioactive element, found among the products of nuclear fission (it's produced when a plutonium isotope decays). Its activity is huge, 127,000 million Bq per gram.

An americium smoke detector is based on the principle that the isotope ^{241}Am emits alpha particles which collide with oxygen and nitrogen atoms in the air, giving rise to ions, electrically charged particles. Applying a low voltage between two electrodes the ions are trapped, and this induces a small electric current. When smoke enters the detector, alpha particles are absorbed by the tiny smoke particles floating in the air and the current is interrupted, making the alarm go off.

A great part of smoke detectors use this dangerous isotope, which was discovered by the Manhattan Project. The reader may have one of them installed above his or her head, continuously sending gamma rays. But there's no need to rush out; the amounts of ^{241}Am used by smoke detectors are exiguous (about a microgram), and the dose you're receiving doesn't exceed 0.001 mSv per year, that's 400 times less than natural radiation. You can relax.

The dose makes the poison, not the substance.

References

Robert, R. (1974). *Marie Curie*, New American Library.
Smil, V. (2006). *Transforming the twentieth century*. Oxford: Oxford University Press.

Chapter 9
Nuclear Reactors

The preference of wildlife for nuclear-waste sites (meaning
Chernobyl) suggests that the best sites for its disposal are the
tropical forests and other habitats in need of a reliable
guardian against their destruction by hungry farmers and
developers.

James Lovelock, "The revenge of Gaia"

Nuclear Energy, A Source of Electricity

It can't be denied that he peaceful use of nuclear energy has always suffered from
the stigma of being associated with warfare. The first nuclear reactor was developed by Fermi and Leo Szilard during the Manhattan Project with the aim of
showing how a fission chain reaction works. The design of the most common
reactor of our days, the LWR, is based on the reactors developed for nuclear
submarines. On the other hand, the initial expansion of nuclear energy went hand
in hand with the growth of the nuclear superpowers. The USA, Russia, France and
the UK simultaneously built power plants in order to generate electricity and
produce a nuclear arsenal. Two processes that can be used both to obtain nuclear
fuel and as a raw material for atomic warheads and are uranium enrichment and
the reprocessing of fuel used in a reactor. Let's see why.

Uranium Enrichment

Most nuclear reactors need fuel with around 3% of ^{235}U. On the other hand, the
classic atomic bomb requires almost pure ^{235}U.

In both cases, it's necessary to separate ^{235}U from ^{238}U in a process known as
enrichment (Fig. 9.1). Uranium is found in many mines all over the Earth—though
Australia alone holds more than 20% of the reserves. After extraction, it is processed
in specialized refineries to form a uranium oxide (U_2O_3) called yellowcake. In the
enrichment plant, the first step consists in transforming the oxide into uranium
hexafluoride (UF_6), which can be gasified at moderate temperatures. The gas then
passes through centrifuges where the isotopes are separated making use of the fact
that their masses differ. Once enriched, UF_6 turns into uranium dioxide, UO_2.

As was to be expected, nuclear powers used their enrichment plants to obtain
^{235}U both for commercial and for war purposes. Some countries that later

J. J. Gómez Cadenas, *The Nuclear Environmentalist*,
DOI: 10.1007/978-88-470-2478-6_9, © Juan José Gómez Cadenas 2012

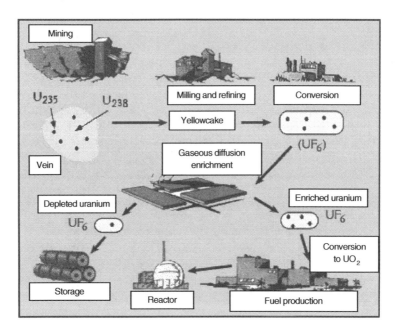

Fig. 9.1 Uranium enrichment

developed nuclear energy exclusively for peaceful purposes (as for example Spain) haven't had the need to develop all the infrastructure, because it's more economical to buy already enriched uranium (in our case, from France). Others, however, couldn't wait to build their own plants: China, India, Israel, South Africa, Pakistan... in all these cases, the final result was developing atomic bombs. Consequently there's little doubt about the real intentions behind Iran's nuclear program, one of the countries with the greatest natural gas reserves on Earth, especially considering how determined they are to control enrichment techniques.

Nuclear fuel is obtained from enriched uranium; this goes into cylinder capsules about the size of a vitamin pill (2.5 cm diameter, 2 cm long) which in turn are placed in long metal tubes (usually zirconium, a hard and corrosion resistant metal, similar to steel), about three and a half meter long and 2.5 cm wide (Fig. 9.2 left). Then the tubes are assembled into *elements* containing about 50–100 units (Fig. 9.2 right). These fuel elements are the basic unit which is fed into the reactor core.

How does a Nuclear Reactor Work?

A nuclear power plant is very similar to a thermal power station fuelled with coal or natural gas. All of them have a large Parsons turbine propelled by a jet of high pressure, high temperature steam. The turbine in turn moves an electrical power

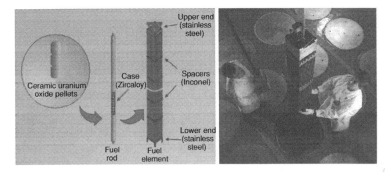

Fig. 9.2 A fuel element

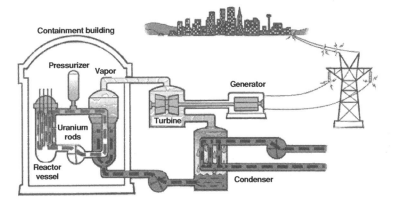

Fig. 9.3 Diagram of a light (pressure) water reactor LWR/PWR

unit. The difference between the conventional plant and the nuclear plant lies in the fuel that is used to obtain the heat necessary to produce steam: in the latter, instead of burning coal or natural gas, a controlled fission reaction is sustained inside the reactor.

Figure 9.3 shows a diagram of the most common reactor nowadays, the *Light Water Reactor (LWR),* often called *Pressure Water Reactor (PWR).* The reactor core is a lattice of around 200 fuel elements assembled in a cylinder about 3.5 m across and 3.5 m high, contained in a pressurized vessel made of steel, with a thickness between 25 and 20 cm.

The vessel is housed in a concrete dome or bunker shielding against radiation, in turn enclosed in a reinforced concrete building, designed to prevent leakage of radioactive gas or fluids in case of accident. The chain reaction is controlled by means of control rods, made of neutron absorbing material, such as cadmium or boron. They can be lifted or lowered into the core. In case of accident, or when refueling, the control rods are lowered completely, so the chain reaction stops.

Apart from the control rods, there are emergency systems allowing to inject a neutron absorber directly into the moderator, in case the chain reaction has to be stopped at once.

The elements containing the fuel are immersed in water circulating around them. In an LWR reactor, the water plays two very different roles. On the one hand, it acts as *refrigerator*, transporting the heat built up due to the nuclear fissions to a secondary loop, as can be seen in Fig. 9.3. In a PWR, the water is pumped into the core at a temperature of 290°C and leaves at around 325°C (without reaching boiling point, because it's under high pressure). Once outside the core it passes through an exchanger, where heat is transferred to a secondary loop that works at a lower pressure and where the steam is generated that drives a powerful turbine connected to a generator. After passing through the turbine, the steam condenses and enters the heater again. It's important to note that the water in the secondary loop doesn't get into contact with the radioactive elements in the core, so the section of the plant dealing with electricity generation needs no special precautions.

The second function of the water in the core is to act as a *moderator*. This is not a fortunate choice of a word, as it seems to imply that the water moderates or tempers the chain reaction. In fact, it's quite the opposite: if the moderator weren't there, the chain reaction wouldn't be possible. Its role consists in moderating the speed of the neutrons emitted during the fission of ^{235}U, not the reaction itself.

Why is it necessary to moderate the speed of the neutrons? One way to understand this is to picture the ^{235}U nucleus like a large drop of water clinging to a glass surface on a rainy evening. Other, much smaller raindrops (the neutrons) slide down close to it. If one of them approaches slowly enough for the big drop to capture it, this results in a drop too large to hold, which splits in two. But if the small drops slide by too quickly, there's no time to capture them and the large drop doesn't split.

If the fuel were made up of pure ^{235}U, the plentiful neutrons going round and round among the uranium nuclei would end up splitting them sooner or later. But let's not forget that *97% of the fuel isn't ^{235}U but ^{238}U*. Let's once again imagine the windowpane on a rainy day with the tiny drops sliding down the glass. But now, for every 3 large ^{235}U drops, you add 97 marbles made of some absorbent material. The little drops, blown by the wind, move very fast and hardly ever stay long enough next to the big ones for these to capture them. However, for the absorbent marbles it's very easy to gobble them up each time they run into each other. As there are many more marbles (^{238}U) than large drops (^{235}U), the quick little drops (that is, the energetic neutrons generated in the fission process) end up being devoured by the absorbent matter and the fission process comes to a halt.

The way to offset this effect is to add a moderator, that is, an agent forcing the neutrons to move slowly (we can picture it as a very thin film of glue which forces the little drops to roll very slowly and makes it possible for them to get into contact with the large drops before being captured by the sponge marble).

The neutrons emitted by the splitting ^{235}U nucleus have a high energy and therefore high speed. If the medium in which they're moving contains light elements (such as hydrogen, helium or carbon), the neutrons hit these nuclei (the glue we pictured) over and over again and gradually lose energy (due to friction) until the become slow neutrons (neutrons with little energy).

We've already seen that the fuel rods are immersed in pressurized water circulating around the enriched uranium capsules, taking away the heat produced by the fission and carrying it to the circuit where steam is produced. Besides, water is the element where the neutrons produced by the fission of the ^{235}U move around. Each liquid molecule contains two hydrogen atoms, that is, two protons, whose task it is to moderate (brake) the neutrons running against them. The slower the neutrons, the easier it is for them to be captured by the ^{235}U, and not be eliminated by the ^{238}U. Therefore, the role of the moderator (in this case the water) is not to moderate, but to enable the chain reaction.

Water is a reasonably good moderator, but not a perfect one, as hydrogen has a tendency to capture neutrons and form deuterium (an isotope of hydrogen with one proton and one neutron) and thus both processes (stopping the protons and deuterium formation) are competing in an LWR. Other light elements, such as carbon (in the form of pure graphite), are much better moderators, but, as we will see, their use entails more risks than using water.

The Composition of Spent Fuel

What happens to the fuel during the three years it remains in the reactor? To start with, ^{235}U gradually disappears, as the nuclei undergo fission. At the same time, the elements created in each fission build up in the fuel pellets. Uranium tends to break up more or less in half, but a great number of isotopes of different elements are produced, among them krypton, strontium, zirconium, molybdenum, technetium, ruthenium, rhodium, palladium, tellurium, iodine, xenon, cesium and barium. Some of these isotopes constitute the notorious "highly radioactive waste". In Chap. 8 we will discuss radioactivity to shed some light on this vague concept.

First of all, as we have seen, the more radioactive isotopes are those who don't last long. Many of these products have half lives (the time it takes a given amount of radioactive matter to be reduced to half) ranging between a few minutes and several weeks, so they disappear after a relatively short time span. Others, on the contrary, have very long half lives. The longer it takes them to decay, the less radiation they emit per gram of matter (the most extreme case is uranium-238). In a way, the most troublesome waste is the waste with a half life long enough to last for quite some time and short enough to emit a lot of radiation per gram of matter.

Besides, not all isotopes are produced in the same amounts. Among the "medium lived" isotopes, the only ones produced in appreciable quantities are cesium-137 and strontium-90, both with half live of 30 years. Among the longer lived isotopes, the most abundant are technetium-99 (half life around 211,000 years) and cesium-135 (2.3 million years).

Finally, the most important point. Not all of them are equally dangerous. Almost all of the elements with a long half life emit only beta radiation, so they are harmless unless it's taken in large amounts (more on this in the next chapter). Not less important is the ease with which an isotope can be absorbed by the human body. Krypton-85, for example, is a noble gas (and thus breathable), its half life is slightly above 10 years (so most of the activity in the sample is lost after a century), and it emits gamma radiation, capable of damaging lung tissue. But sometimes these noxious effects can be turned to our advantage. Strontium is an element that can be assimilated into our bones (it has the ability to "imitate" calcium chemically), and based on this, strontium-89 is used in radiotherapy to fight bone cancer.

When we take the fuel rods out of the reactor, 94% of the pellets consist of uranium in which the excess ^{235}U has undergone fission (so it's quite similar to the original, natural uranium). 5% consists of products of ^{235}U fission, many of which have very short half-lives and thus lose their activity after a few years. One decade later, activity is dominated by ^{137}Cs and ^{90}Sr, together making up 0.3% of the spent fuel. In the end, the fuel contains 0.9% plutonium and 0.1% of other trans-uranians such as plutonium-230 end americium-241.

Balance of materials to fuel a 1,000 MW nuclear power station during a year	
Mine	20,000 tons of ore (1% U)
Refining	230 tons of yellow cake 195 tU
Conversion	288 tons of UF_6 195 tU
Enrichment	35 t of UF_6 24 tU, enriched
Fuel	27 t of UO_2 24 tU, enriched
Spent fuel	27 tons 23 tU, 240 kg Pu, 720 kg fission products

The table shows the data for a typical reactor with a power of 1,000 MW. From 20,000 tons of uranium ore you get 24 tons of enriched uranium, enough to feed the plant for one year and obtain around 7,500 GWh electricity. The spent fuel contains 23 tons of uranium in which the rate of ^{235}U has gone down to 0.9% (that's a bit more than in natural uranium), 240 kg plutonium and 720 kg fission products.

In Fig. 9.4 we can see how the activity of the waste evolves over time from the moment the fuel is taken out of the reactor. The horizontal line stands for natural radioactivity of the equivalent amount of uranium ore, and it allows us to judge when the waste is not more dangerous than a natural uranium vein. The fission products (^{137}Cs and ^{90}Sr) dominate global radioactivity during the first 100 years, but after three or four centuries their radioactivity level has fallen below the natural radioactivity in the original fuel. However, the lower but more persistent contribution from trans-uranians (dominated by ^{241}Am) is added to radioactivity stemming from fission products, so global radioactivity remains above radioactivity produced by uranium for about 10,000 years.

When the fuel is taken out of the reactor, its activity is extremely high and therefore generates a lot of heat, due to gamma radiation. This is the reason why it is stored in pools in the nuclear plant itself. The water stops gamma radiation and

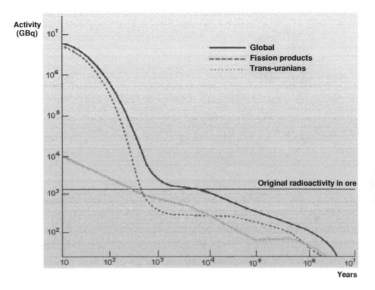

Fig. 9.4 Variation in the radioactivity of waste over time. *Source* (WNA 2007)

has a cooling effect. It's important to note that gamma radiation, though very intense, has a short range. A few meters of water are enough to shield it: outside the armored building where the pools are housed, radioactivity doesn't exceed normal levels.

After around ten years, the short-lived isotopes have disappeared, the activity of the waste has been reduced to one hundredth, and the fuel bars can be taken out to be stored or reprocessed, though they are often kept in the pool for a few more decades, a "temporary" solution which in Spain is still in place. Our pools are now at 75% capacity and will be full in around ten years, so they will have to be shut down soon.

So what can we do with them? There's three possible approaches: (a) reprocessing, (b) temporary dry cask storage, sealing the waste into steel cylinders shown in Fig. 9.5, (c) vitrification for temporary or permanent storage in a geological deposit.

The last two options are related to waste storage, which will be dealt with in Chap. 10. Let's now turn to the possibility—or, in my opinion, necessity—of reprocessing them.

Waste Reprocessing

Why should spent fuel be reprocessed? To start with, because 23 tons of the 24 tons we have taken out of the reactor are natural uranium, which can be reused as fuel or stored in deposits that don't require special safety measures, as radioactivity is very

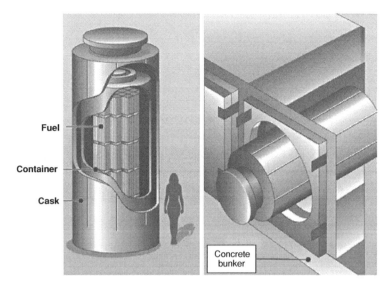

Fig. 9.5 Dry cask storage containers for temporary waste storage. *Source* (WNA 2007)

low. Concerning the 240 kg of plutonium, there's no reason not to use it as a fuel. In fact, in a conventional reactor, almost 30% of the electricity is generated thanks to fissions of the plutonium that has accumulated from decaying ^{238}U. This leads to the idea of processing spent fuel, separating genuine fission waste (just 5%) from uranium and plutonium. The advantages appear obvious:

1. A more economical use is made from the fuel, getting an extra 25% energy from uranium and plutonium
2. The amount of waste to be stored in safe geological deposits is reduced enormously (less than a ton per year in each nuclear power station, compared with 24 tons per year and power station).

The standard method for processing spent fuel is PUREX, an acronym standing for *Plutonium and Uranium Recovery by Extraction*. The first step is dissolving the fuel in nitric acid. Then an organic solvent, a mixture of 30% TBP and 70% kerosene, is used to extract the uranium and plutonium, while the other fission products remain in the nitric acid. In the second step, uranium is separated from plutonium. Uranium can be stored or used to obtain newly enriched uranium, and plutonium is the base of a fuel called MOX (an oxide mix of plutonium and uranium). Finally, the 720 kg of radioactive fission products are vitrified and transported to a permanent deposit.

Several countries, among them France, the UK, the USA and Russia, built reprocessing plants in the 1970s. France still reprocesses fuel, as does Japan, which has recently built its own installations. In the USA, however, civil reprocessing was abandoned in 1977.

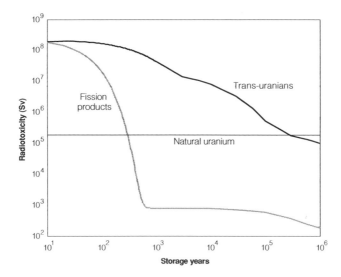

Fig. 9.6 Radiotoxicity (RT) of spent fuel compared to RT of natural uranium. The most dangerous elements are trans-uranians. *Source* (WNA 2007)

Is there some reason no to reprocess the fuel, in view of the obvious advantages of reusing plutonium and reducing the waste volume? Unfortunately, one of these reasons—the most important in my opinion—is economy. Uranium is very cheap. At less than a hundred dollar a kilogram, reprocessing fuel is three times as costly as using fuel only once. Consequently, the nuclear industry isn't very motivated to take on the complications associated to the "closed cycle" (recycling and reprocessing of elements). It pays off to stick to the "open cycle" (only one use), especially if, as is the case in Spain, the problem of waste storage is put off by leaving it in pools for decades.

I believe that the economic argument should not carry a greater weight than ecological considerations. Figure 9.6 shows radio toxicity (RT) of fission products compared to natural uranium. RT is measured in unit doses, which specify the potential damage to the organism (Chap. 10), but what is relevant here is the comparison with the level of natural uranium. The trans-uranians are much more dangerous, and remain so for much longer, than other fission products, like ^{137}Cs. The most dangerous of all is plutonium, whose radioactivity level doesn't fall below natural uranium until a 100,000 years have passed, followed by americium (10,000 years).

The obvious conclusion is that fuel has to be reprocessed, not only in order to make better use of the energy and reduce the amount of waste to be stored, but above all to get rid of the more dangerous, longer lasting waste. In spite of geological deposits being safe (Chap. 10), if we don't bury the plutonium (and other trans-uranians that also need to be separated from the fission products), the time until they cease to be dangerous is reduced from 10,000 years to three or four centuries.

Plutonium's Black Legend

But then things get complicated when other considerations come into play, related to the potentially destructive uses of plutonium. It was these considerations that led the Carter administration to close down the US reprocessing plants in the 1970s.

Plutonium is of course suitable for making bombs, and PUREX was used by all great nuclear powers to separate this material in order to build nuclear warheads. In the 1960s and 1970s, the same powers, with the USA at the forefront, started to fear that a great number of other countries might get their hands on nuclear arms (but that didn't detract them from making their own). That was the origin of the "problem of nuclear proliferation", which could be expressed like this: A country interested in developing the bomb pretends its only motivation is to generate electricity. It builds a nuclear reactor and after a year holds 250 kg of plutonium. It sets up a plant to reprocess the fuel used in the reactor along the year, separates the plutonium and is capable of making a few nuclear warheads. Sounds familiar? We're talking about the 1970s and it might refer to the current situation in Iran. An irresistible story for any fiction writer[1].

Why did the USA put an end to civil reprocessing of fuel? To prevent the risk we have just described. If a country doesn't have plants to separate plutonium, it can't make any atomic weapons even if it has nuclear reactors. That's why the "political" solution to the problem of proliferation was—according to the Carter administration—to close down all civil reprocessing plants, starting in their own country to set a "moral example" and be able to exert pressure on the other partners.

Incidentally, the morality of the example is subject to debate, because the USA never closed down their military reprocessing plants, where they continued to use PUREX to separate plutonium and build tens of thousands of nuclear warheads. And anyway it was of no use, as other nuclear powers, France among them, paid no attention to their powerful partners' guidelines.

In fact, the problem of nuclear proliferation is one more example of how peaceful applications of nuclear energy are held accountable for sins they haven't committed. If you study attentively the argument about the country that builds a nuclear reactor and a reprocessing plant with the only purpose of making a bomb, you see it's almost ridiculous.

Why? First of all, the lifecycle of a civil reactor is about one year. Every 12 months the core is opened and two thirds of the fuel are replaced. Of course in this third there's more than enough plutonium (about 90 kg), but in one year you don't only get ^{239}Pu (from ^{238}U) but also plutonium-240 (from ^{239}U), which makes up approximately 10% of the total plutonium.

Let's recall the basic mechanism of the bomb. Two pieces of plutonium of subcritical mass are suddenly brought together and the chain reaction, which cannot be sustained in either of the two pieces separately, goes off when the critical mass is exceeded and the number of neutrons grows exponentially. And what happens if we bombard the two pieces of plutonium with neutrons before we join them? A lot of

---

[1]   Including yours truly. See http://www.espasa.com/Materia-extraña551516.

fissions take place, but there's no explosion as the critical mass isn't reached; plutonium is used up and we are left with something like a wet firecracker.

But then, plutonium-240 is a prolific neutron emitter, and 10% of it added to ^{239}Pu is enough to make the explosive go bad. The longer we wait until we open the reactor, the more plutonium-240 is formed. The only alternative is to open it after a month or two, in which case the plutonium mass is already too small for our purposes. Then there's the little problem of explaining to the International Atomic Energy Agency[2] why we have shut down the plant for one month to take out the fuel we had just started to use.

Fact or fiction? And it appears almost ridiculous when we realize that, if our purpose is producing plutonium, we don't need a great 1,000 MW reactor at all.

It's much more convenient to go for a reactor designed exclusively to produce plutonium, which, among other advantages, can work in low pressure and low temperature conditions, and uses natural uranium (we only intend to transform ^{238}U into ^{239}Pu, without generating electricity). We can get a device like this at the tenth of the price of a commercial reactor, and build it much faster. Naturally, all the plutonium used for nuclear warheads in the USA or France was obtained in military installations. The Russian reactors were the only exception, and this led to an unsafe design that caused the Chernobyl accident (See this chapter).

Let's now look at the reprocessing plant. Why should we build a complex, expensive facility when a small dedicated factory, like the military installations the USA never abandoned, is sufficient? If a country really wants to make an atomic bomb, it's much easier to resort to military facilities—clandestine, if need be— than trying to fool the IAEA inspectors.

I think that a lot of the people who oppose nuclear energy mistake the instrument for its use. The day the knife was invented, humans had found both a tool and a weapon. Banning loggers from cutting down trees with axes for fear that terrorists might get hold of these axes to cut off heads is a bit naïve. Most of the time, it's easier for violent people to craft their own weapons rather than use other people's peaceful tools.

Plutonium can be used to make bombs and to generate electrical power. Banning the reprocessing of fuel in a commercial reactor seems to me a strategy similar to banning loggers' axes. Whoever wants to get his hand on plutonium can resort to much more efficient, safer, cheaper and more inconspicuous means to obtain it than stealing it from nuclear plants.

In any case, there are alternatives to PUREX that get rid of the problem. UREX, for example, allows to separate uranium from all the other trans-uranians. Following similar arguments to the ones we have seen, if ^{239}Pu is mixed up with these other isotopes, it can't be used to build a bomb, but is still useful as fuel for a reactor. Many of the trans-uranians can also be "burnt" (that is, split) in special nuclear reactors, as we will see shortly.

---

[2] The IAEA seeks to promote the peaceful use of nuclear energy, and to inhibit its use for any military purpose. It sets up rules concerning nuclear safety and environmental protection, helps member states by means of technical cooperation and encourages the exchange of scientific and technical information about nuclear energy.

Another risk that is often mentioned is the possibility that a terrorist group might get hold of enough plutonium to build an atomic bomb. In practice, however, plutonium is one of the most strictly controlled substances in the world, and as we can unfortunately see day to day, there are much more viable and economical ways to commit terrorist attacks. Let it be said, by the way, that the technology which is necessary to build a bomb is not within everybody's reach: even a "homemade" bomb needs a team of several nuclear physicists and engineers, plenty of technicians plus the appropriate facilities. It's much easier to turn to dynamite.

How does the risk of a terrorist group obtaining plutonium—and learning to build a bomb—compare to other similar threats? Is it more or less difficult than developing a deadly mutant flu virus and spreading it in public spaces all over Europe? More or less probable than a cyber attack using computer viruses to poison the water supply? Or than bursting a reservoir dam and causing a giant flood in the neighboring town? Would a low power homemade nuclear bomb kill more people than a few tons of easily obtainable napalm, released in a city centre at rush hour? As any techno catastrophe film writer can tell you, the list is never-ending, and the damage many of these potential attacks can cause comes close, or even exceeds—in the case of a lethal epidemic—the effects of a nuclear explosion.

Terrorism is one of the great evils of the 21st century, and to fight it needs much more than getting obsessed with unlikely scenarios. Up to the moment, the fears related to plutonium and civil reactors have proven to be quite unjustified; in fact, the opposite is happening. A great amount of nuclear arms are being dismantled, its plutonium turned into fuel to generate electricity. Even that element whose name invokes Dante's inferno can turn from sword to plough. It's up to us.

Paracelsus and the Rose

In one of Jorge Luis Borges' most beautiful stories Paracelsus, the great alchemist, is sitting at home, looking at the fire. He is very old and doesn't want to die without passing his knowledge on to someone. He asks God—any god—to send him a disciple. Shortly after, there's a knock at his door. A young, polite man steps into his hut and asks for permission to follow him on his path to wisdom. But Paracelsus joy won't last long. The youth, surprised at not seeing the test tubes and retorts he expected—the old man's home is as empty as a monk's cell—asks for a proof. He is soft-spoken and his manners are exquisite, but he doesn't trust the old man.

To the wise man's dismay, the youth gets hold of a rose which brightens the humble table and throws it into the fire. Once consumed, he asks the alchemist to make it rise again from its ashes. Paracelsus replies he is unable to perform the miracle and the youth goes away, embarrassed and relieved to have found out in time about the old charlatan's fraud.

The story reminds me of the predicament of the peaceful applications of nuclear energy. The alchemist—represented by the scientists who believe this is a con-centrated, clean and plentiful energy source—is questioned by the incredulous youth, who demands more and more proofs without ever being satisfied and finally throws the flower into the fire and asks for a miracle.

It's true that, the same as with Paracelsus's would-be disciple, his mistrust is not unfounded. The former can't believe that an old man without a distiller is the wizard who says he has found the philosopher's stone that transforms lead into gold. The latter mix undeniable facts—obtaining energy from the atom is a complex process requiring large investments, sophisticated technology and strict safety protocols—with unfounded apprehensions, many of which are derived, as I have pointed out several times, from the long shadow thrown by the warfare applications of this technology.

For the last decades, the critics of nuclear energy have opposed waste reprocessing for fear of atomic arms proliferation and have at the same time opposed the storage of waste in geological deposits, on the grounds, among other things, that plutonium might be stolen for terrorist purposes. It's a double bind situation for nuclear energy. By blocking all solutions, it is shown—by definition—that plutonium is an "unsolved problem". The plutonium that is stored in temporary or geological deposits comes from fuel that has spent three years in a reactor and contains a high proportion of plutonium-140 and other trans-uranians which render it essentially useless for arms manufacture.

Nuclear Alchemy: Fast Nuclear Reactors

When the aspiring disciple throws the rose into the fire and asks the old alchemist to make it rise again, the wise man declares himself incapable of doing so. How could he? It's not and easy miracle to perform. And yet, something similar happens inside a kind of nuclear reactor called Fast Nuclear Reactor, FNR.

Instead of uranium enriched to 3% ^{235}U, a FNR uses MOX (obtained form plutonium) as fissile fuel. The MOX elements are surrounded by a layer of depleted uranium (with a lower content of ^{235}U than natural uranium). The core of the FNR houses, quite logically, a refrigerator to transfer heat to the turbine and a *reflector*, that is, a material surrounding the fuel whose purpose is to "reflect" the neutrons that escape from the core (and would be absorbed by the massive concrete walls of the bunker) back to the reaction. In exchange, a FNR doesn't use a moderator. As its name implies it works with fast neutrons, capable of splitting the ^{239}Pu that replaces ^{235}U as a fuel.

Besides, these fast neutrons are especially apt to transform the ^{238}U of the outer layer into ^{239}Pu. This also happens, as we know, in a conventional reactor, but in a FNR the process is more marked, because fast neutrons are more easily captured by ^{238}U.

And so? Well, when we recover the spent fuel we find that we *have produced more plutonium* (by uranium transmutation) than fission has consumed.

Paracelsus's dream has come true. Nuclear physics acts like a philosopher's stone capable of transmuting lead (that is, the ^{238}U which can't be split and thus is useless for energy generation) into ^{239}Pu, energy producing gold.

This philosopher's stone enables us to use all of the uranium as a fuel (transforming it into plutonium), not just the ^{235}U, and this means the ore reserves are multiplied by 100 (currently we use less than one per cent of the uranium we mine). As we will see, *cheap* uranium reserves are considered to be sufficient to feed conventional reactors for about a century—though they're probably much greater and, at a price, in fact inexhaustible. Therefore, using FNR there would be enough fuel for $100 \times 100 = 10,000$ years. That's the time span that separates the end of the last ice age from the first agricultural settlements.

How to Get Rid of Highly Radioactive Waste

By using FNRs, uranium and plutonium are used up completely, so the radioactive ashes consist only of fission products (less than a ton per years for a 1,000 MW plant, whose activity is harmless after 200 or 300 years) and the rest of trans-uranians, not more than 30 or 40 kg per year, but whose contribution to radio toxicity (in the case of ^{241}Am) is meaningful. It's a bit frustrating that less than 50 kg materials per working year make us develop the complex protocol of a geological deposit certified for ten thousand years.

A solution to the problem, which incidentally would appease the proliferation worries, is to replace PUREX by new reprocessing techniques such as UREX or the interesting pyrometallurgical techniques that are being developed in the USA (see for example Hannun et al. 2005) and allow to separate, on one side, uranium, and on the other, *all trans-uranians* (including plutonium) and finally fission products. This way we would be killing two birds with one stone.

1. ^{239}U is not isolated from the rest of the trans-uranians, but is polluted with the other isotopes, so there is no risk that it might be used as a weapon.
2. We can produce a new kind of fuel that mixes uranium with all the trans-uranians (and not just with ^{239}Pu). Fast neutrons are able to split most of these isotopes. So the FNR can be used as a "waste burner", destroying the ones with a longer half-life.

With this strategy, resources are made a rational use of, as opposed to the squander of the open cycle; the risks of proliferation are virtually eliminated and the trans-uranians with long half-lives are no longer present in the final ashes. As a result, waste loses most of its activity in a few centuries, which on the other hand removes the problems related to its storage.

The Phoenix Bird

The idea behind the fast neutron reactors isn't new, and in fact great sums have been invested in their development, without achieving a complete success up to now.

FNRs are more difficult to operate than conventional reactors, in part because the reaction with fast neutrons can run out of control more quickly, in part because the most appropriate coolant—liquid sodium—reacts violently with water and air and needs special safety measures. These technical complexities aren't insurmountable, but they make FNRs more expensive compared to LWRs. On top of it, fuel reprocessing is costly. Nuclear industry isn't different from petrol industry or any other commercial activity in that it strives to reap economic benefits. At present, it comes cheaper to gobble the oil from the Near East fields rather than extract if from the tar sands in Canada; similarly, it's more economical to use the open cycle and conventional reactors rather than invest in more costly and complex FNRs.

The development of FNRs began in the 1970s, when developed countries, strangled by the OPEC, where considering the need to abandon fossil fuels more seriously than today. At the time there was also the fear that the input material (particularly uranium) might become scarce at any moment (more on this in Chap. 11). Super phoenix, one of the most important fast neutrons reactors of the last decades, was approved in 1972, in full oil crisis.

Unfortunately, the reactor wasn't completed until 1984 and the construction costs were high. In a way, it's not surprising, because the technology was more complex than the one associated with conventional reactors, which worked perfectly in the French nuclear scheme by repeating the same idea 56 times. But we can't deny that the nuclear industry has often been too optimistic. In hindsight, it would have been better to achieve a detailed understanding of the problems associated with operating the FNRs, by means of intense R&D, like it's now carried out for the so-called Generation IV Reactors, of which more later. But in those days it seemed that fast neutron reactors and "plutonium economy" were good business. However, by the time the reactor was finished, this wasn't the case anymore. Oil prices had fallen again, it became increasingly clear that there was no lack of uranium, and to make matters worse, green opponents of atomic energy had turned Super phoenix into public enemy number one, including—in 1982—one of the few recorded terrorist attacks against a nuclear reactor. The author was Cha Nissim, who in 1985 was elected to a MP office at the cantonal parliament of Geneva, under the aegis of the Green Party of Switzerland.

Super phoenix was put into operation and reached over 90% of its nominal power, most technical problems having been solved, but was finally shut down in 1998 after many lawsuits filed by eco-activist organizations. Paradoxically, in a country were nuclear energy was and still is widely accepted, the battle wrought against Super phoenix by anti nuclear groups was fierce and in the end successful. The French government chose to offer the reactor as propitiatory victim and let things calm down.

In brief: unripe technology, mistaken economic calculations and green opposition threw the philosopher's stone overboard. The rose was thrown into the fire and burnt up.

The Future of Nuclear Energy

As we have seen, the history of nuclear reactors began more than sixty years ago with the first prototype developed by the Manhattan Project. Between 1945 and 1970 several reactors were built—mainly in the USA, Canada and Russia—which were based on natural uranium and moderated by graphite, similar to Fermi's first reactor. These reactors have been called Generation I of nuclear industry. Unfortunately, the most well known of these is the one at Chernobyl.

From 1970 to the mid 90s nuclear energy expanded considerably: more than 450 reactors were built, almost all of them LWRs, operating with enriched uranium and water as a moderator. The Canadian reactors (natural uranium, heavy water as a moderator) were an exception, together with a few others. This was Generation II.

Generation III refers to the reactors that are now being built or projected. They're essentially enhanced versions (safer and more economical) of Generation II reactors.

Generation IV is the name given to an R&D program signed by ten countries plus EURATOM (European Atomic Energy Community) with the aim of producing reactors that are cheaper, safer and more flexible, and also capable of sustaining a closed cycle (that is, FNRs). The goals of Generation IV are: achieving a "sustainable cycle" (in other words, getting rid of trans-uranians and making us of all the uranium, for which you need FNRs) and enabling the applications of nuclear energy for the future hydrogen economy. They also intend to enhance the already excellent safety of Generation III and to reduce costs—by introducing standardized designs that allow mass production of reactors—so that they can compete with coal fired power stations.

When will we see these new improved reactors? Probably not before 2030, which means the reactors to be built in the next 20 years are likely to still be Generation III. This implies that the fuel used in the next two decades—which can be cooled in the reactor pools after going through the reactor—could be recycled and used directly by Generation IV reactors, which would extract the 95% remaining energy in the uranium, at the same time doing away with trans-uranians and thus the problem of storing waste.

The decades of anti nuclear stubbornness—which have been suspiciously coincidental with the years of cheap fuel and ecological unconsciousness, at least in relation with climate change, an open secret for decades—may have had a positive side. The nuclear industry has often acted hastily, embarking on ill-planned projects whose final cost was higher than foreseen, or stumbling over problems that could have been solved with more R&D before initiating commercial applications. Generation IV may be the first whose concepts have been filtered by research over three long decades before going commercial. That's the way to do things.

Natural Reactors

One of the most stubborn superstitions I often hear portrays renewable energies as "natural" sources, contrasting it with nuclear energy, an anti natural product of human technology.

And yet, about 2,000 million years ago, Gaia carried out a complete experiment on her own account.[3]

It's a story you can't miss. In May 1972, a worker at a reprocessing plant for nuclear waste in France spotted a suspect phenomenon while routinely checking some apparently ordinary uranium. The rate of ^{235}U in natural uranium has been measured in mines all over the Earth and it's always the same, exactly 0.720%. But in the sample we're referring to, mined in Oklo, Gabon (a former French colony in central Africa), ^{235}U only made up 0.717%.

A bit more than 4 parts in 1,000 doesn't look like a great disagreement, but it was enough for the French scientists to suspect that something strange was going on. In other parts of the mine, as it was later shown, the uranium ore contained an even lower rate of the fissile isotope. All in all, about 200 kg of ^{235}U was *missing*.

In fact, the answer to the riddle had been anticipated by the American physicists George W. Wetherill and Mark G. Ingham, who in 1952 suggested that certain uranium deposits might in the past have acted like natural versions of fission reactors.

In this chapter we have seen what is needed to build a nuclear reactor: enriched uranium, containing approximately 3% of ^{235}U, a moderator to brake the neutrons produced during fission (water is a good choice) and control rods made of born or other materials tending to capture neutrons, in order to stop the chain reaction if necessary.

The half life of ^{235}U is about 700 million years, which means that every 700 million years the amount of ^{235}U on our planet is reduced by half. Therefore, 2,000 million years ago there was much more of it, natural uranium couldn't contain just 3% of this isotope. A civilization inhabiting the world at that time wouldn't have needed to enrich uranium in order to make their nuclear reactors work.

Nor did Gaia. But she did need a moderator and a region without natural "control rods", that is, a region free of boron, lithium and other elements with an appetite for neutrons.

In the Oklo region, a watercourse leaking into an uranium mine acted as a moderator. The archeological evidence points to reactors working for several million years, until the ^{235}U ran out, or rather, up to the moment when there was too little of it to sustain the chain reaction.

Once the reactors went out, the mine turned into a radioactive waste deposit. There's still enough plutonium left to prove that the materials have remained in place, buried under granite and soil, for the last 2,000 million years, in spite of the

---

[3] See for example http://www.ocrwm.doe.gov/factsheets/doeymp0010.shtml.

abundance of water in the region (which was essential to sustain the reaction). You could say that Gaia, besides inventing nuclear energy thousands of millions of years before the appearance of man, solved the problem of how to store radioactive waste, which will be dealt with in the next chapter.

References

Hannun, W. H., Marsh, G. E., & Stanford, G. S. (2005). *Smarter use of nuclear waste*. New York: Scientific American.
WNA. (2007). World nuclear association. http://www.worldnuclear.org/info/inf75.html.

Chapter 10
Nuclear Power, No Thanks?

Some experts fear that the risk of operating the LHC (Large Hadron Collider, a great proton accelerator in CERN, the European Organization for Nuclear Research) disproportionately outweighs anything science might gain from this experiment. ....even CERN scientists concede that there is a real possibility of creating destructive theoretical anomalies such as miniature black holes. These events have the potential to fundamentally alter matter and destroy our planet.

Citizens' website against the Large Hadron Collider
http://lhcdefense.org/

A Hypochondriac Society

The greatest misunderstanding related to nuclear energy, which is continuously exploited by its opponents, is the nature of radioactivity. Pierre and Marie Curie carried in the pockets of their laboratory coats test tubes full of highly radioactive substances, unaware that gamma radiation could be harmful at high doses. In contrast, antinuclear propaganda has been so effective that most people are convinced that any radioactive "leakage" is deadly or that radioactive waste has to be buried at a depth of kilometers to avoid harmful effects—in fact, a few meters of soil is enough to absorb even the most intense gamma radiation—or that cancer risk grows exponentially due to nuclear power plants—which, fortunately enough for French citizens, isn't true.

Cancer. It suffices to speak out the word in a society like ours, and everybody starts trembling like children in olden times when the bogeyman was mentioned. It's true there are good reasons to fear it, as it kills one in every four people in rich countries. In the poor ones, it isn't so successful, having to compete with malaria, typhus, pollution, AIDS, malnutrition or the pure and simple violence of a miserable life. The market is much less competitive in the Western world, where business is shared only with heart disease and traffic accidents. But we are more frightened by cancer; we seem to think that it's the neighbor who will suffer from a heart attack or die in a car crash. Cancer… what can we do to be spared? Stop smoking? (of course). Avoid overeating and drinking moderately? (very possibly). Eating only macrobiotic food? (well, it probably won't hurt). Avoiding saccharine in your coffee? (useless if you take sugar instead). What about electromagnetic waves? (if you browse what has been published in the last 30 years, you will see the pendulum has swung from causing cancer to preventing it and again to causing it). And what about fertilizers, pesticides, preservatives, mobile telephones, TV screens, car exhausts, asbestos, titanium? And as if that weren't enough, there's radiation.

J. J. Gómez Cadenas, *The Nuclear Environmentalist*,
DOI: 10.1007/978-88-470-2478-6_10, © Juan José Gómez Cadenas 2012

Let's picture, for example, a poor devil forced to remain for 1 year in the immediate vicinity of a nuclear power plant. Like Prof. Bernard Cohen explains in his excellent book (Cohen 1990), the poor guy is going to be bombarded by nothing less than 500 billons of high-energy particle during these fateful months, at the impressive rate of 15,000 particles per second.

You wouldn't like to be in his shoes, would you? Unfortunately, we all are. The astronomical number of murderous particles hitting our unfortunate guinea pig is the same any of us is receiving, corresponding to natural radioactivity; the radioactivity level close to a power plant is indistinguishable from this natural radiation. Natural radioactivity comes from different sources: cosmic rays—high energy particles from outer space that pierce through the atmosphere, which acts a partial shield; trace amounts of radioactive materials all around us—for example in granite, brick and many other materials; and natural emissions of radon gas.

Wait a minute. Five hundred billion particles a year? Shouldn't we all be dying from cancer? That's not happening, as we have seen, and the reason is very simple: the probability of a high-energy particle causing cancer is one in 30,000,000,000,000,000! (Cohen 1990).

Cancer is a Russian roulette we keep playing all our lives. The dice aren't just radiation but countless physical, chemical and biological processes. Their probabilities are all very low, and radiation is among the ones with the lowest likelihood. Which, by the way, isn't surprising at all. We have evolved for this to happen.

Though, why should we run any risks? Radiation from nuclear power plants may be low, but it's still an extra dose, isn't it?

It depends. The dose you receive is highly dependent on whether you live in a house made of wood or brick, cement or granite (materials that contain traces of radioactive substances such as uranium, thorium and potassium), increases considerably if you often travel by plane—where we are exposed to cosmic rays that haven't yet been cushioned by the atmosphere—and shoots up when your dentist takes an X-ray of your bad tooth. If you want to play it safe, I'd recommend not climbing any mountains, no skiing trips—radiation increases with height—never going to the beach—ultraviolet radiation roasts your skin—and perhaps installing a lead enclosure around your bed. It helps to sleep unaccompanied, as humans are rather radioactive due to decay of unstable isotopes of calcium and potassium in our bones.

Or we can surrender to the idea that living is a high-risk activity.

What Radiation Dose Do We Receive?

To get an idea of the risk that radiation entails for human health, we need a quantitative measure of exposure. It's important to distinguish between units that measure the activity of a radioactive substance and those that measure its effect on the human body (effective dose). As we saw in Chap. 8, the unit for activity is the Becquerel, which stands for one fission per second. The unit for measuring the effective dose (which

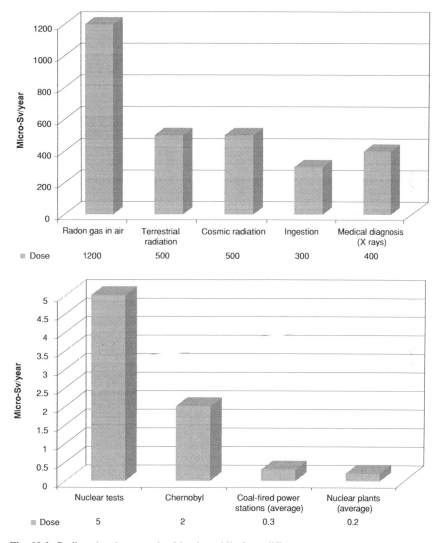

Fig. 10.1 Radioactive doses received by the public from different sources. *Source* (UNSCEAR 2000)

includes the effect on biological tissue) is Sievert or Sv. Frequently used multiples are the millisievert (1 mSv = 0.001 Sv) and microsievert (1 μSv = 0.000001 Sv).

Figure 10.1 shows the doses corresponding to different radioactive sources,[1] expressed in μSv (one millionth of a Sievert). Notice that the scale in the graph below

---

[1] The main source of information for this chapter is a study published by UNSCEAR, the UN Scientific Committee on the Effects of Atomic Radiation, possibly the most reliable and authoritative source on these effects. The study is online (UNSCEAR 2000).

is much smaller than in the upper graph. The dose we receive due to natural radiation is 2,400 μSv per year; the only source of human origin which is not completely negligible, though much lower than natural radiation, is medical diagnosis, that is, X-rays (the global average is 400 μSv per year, but in rich countries, with enhanced medical care, the per capita average is around 1,200 μSv per year). The average doses related to the mortal sins of atomic energy (nuclear tests in the 60, 70 and 80s, Chernobyl accident) are, as can be seen, ridiculously low. It's also noteworthy that *the average dose due to nuclear power plants is lower than the one coming from coal-fired power plants* (which emit a small amount of uranium and thorium, present in traces in the coal, of which tons and tons are burnt). In any case, both are negligible compared with the effects due to habitually watching TV, carrying a light-emitting watch or occasionally traveling by plane.

On the other hand, Fig. 10.2 shows the doses received by people employed in the nuclear industry (mining, operators, educators, etc.) compared with other workers. Curiously enough, doses are much higher among aircrews (owing to their continuous exposure to cosmic rays) or in mines other than uranium mines (because safety measures are less strict, but all metals contain radioactive traces). The most important source remains radon gas, which stems from natural sources, and has more effects at high altitude.

Radiation and Cancer

Let's look at a concrete example. Who would like to live in the neighborhood of a nuclear power plant? According to "green" propaganda, this would almost amount to suicide from cancer. Let's analyze the facts. According to (UNSCEAR 2000), the extra radiation received annually by the suicidal freak who can't be dissuaded from pitching his tent next to the entrance of a nuclear power plant is around 30 μSv. How do his or her odds to contract cancer increase?

Actuaries, Smokers and Loss of Life Expectancy

Some background information is needed to get a clear understanding. We'll have a quick look at the fundamentals of a branch of mathematics called risk assessment. Assurance companies hire professionals called "actuaries" to calculate the price of the insurance policies depending on the likelihood of the accident, the disease or the calamity insured against.

A basic example: every year, about 3,500 people die in traffic accidents in Spain. There are around 20 million cars on our highways. Consequently, a Spanish citizen using this means of transport has, on average, a probability of 3,500/20,000,000 = 0.000175, that is, almost two in 10,000, or one in 5,000 of perishing on the road.

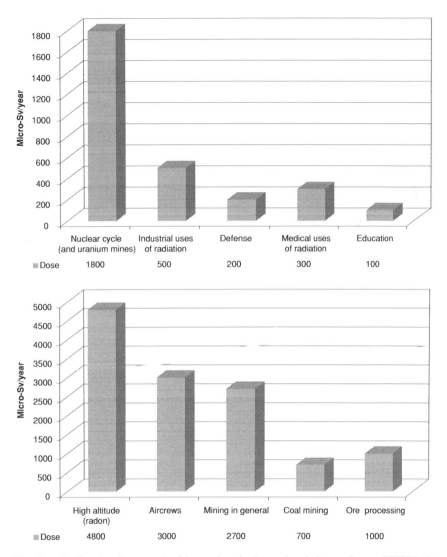

Fig. 10.2 Radioactive doses received by workers in the nuclear industry. *Source* (UNSCEAR 2000)

Obviously, average probabilities are very gross figures. The chances of a road accident aren't the same for a cautious sixty-year-old who drives her car once a month as for the youngster who has just bought a 150 hp motorcycle to run joy-races with his friends. The simple calculation we have done is not enough, and that's the reason why insurance companies have to pay their mathematicians high salaries. And yet, an average figure like the one yielded by this division can help us get an idea of how risky a certain activity is, at least in comparison with others.

Some more examples: the likelihood of dying from a flue is similar to the likelihood of dying in a car accident in Spain (one in 5,000), but lower than the likelihood of dying from a violent death (adding up a lot of possible causes, this amounts to around one in 3,000). The probability of passing away due to leukemia is one in 12,500. If you really want to be in for a short life, there's nothing like smoking 10 cigarettes a day (a chance of one in 200 to do you in). The likelihood of radioactivity dispatching you *if you work in the nuclear industry* is one in 57,000, more or less the same as the likelihood of being wiped out by a tornado, lower than the likelihood of dying in an occupational accident (one in 10,000) and much lower than the likelihood of being annihilated by an earthquake in certain parts of the world (one in 22,000 in Iran) or of being liquidated by an asteroid (one in 20,000, not a negligible figure).

All these probabilities can be encapsulated in a figure called LLE, in honor of its discoverer, Lenny Lewis-Engelman[2]. At the time of his crucial discovery, Lewis-Engelman, an actuary employed in an important insurance company, was forty years old, enjoyed good health and was very successful in life, so he was absolutely determined not to run any unnecessary risks.

Everything started they day Lenny, a moderate smoker, set out to ascertain how risky his habit was. He began by calculating his life expectancy, which turned out to be some 40 years more. Then he did some research into the number of deaths related to smoking in his country—most of them due to lung cancer—and divided them into the total population (the same calculation we have done in the case of road accidents); he found that the likelihood of dying in a certain year because of smoking half a pack of cigarettes was one in 200. Therefore, he deduced, the likelihood of his vice dispatching him in the next 40 years was $40/200 = 1/4$, and this, multiplied by the four decades he still expected to live, yielded the terrifying amount of only $40 \times 1/4 = 10$ years. If he reduced his daily consumption to just one miserly cigarette, he would still be throwing out 1 year of LLE. And if he smoked one cigarette a year, this fancy would still cost him a day of life expectancy.

Just a brief digression. A LLE of 1 day for each cigarette doesn't mean there's a kind of cosmic register that deducts a 24 h period for every cigarette you light. It's just a (very graphic) way of expressing the fact that no cigarette is harmless. If we compare a large smoking population with another large non-smoking population (everything else being equal), the non-smokers live on average ten years longer than the people who smoke ten cigarettes a day.[3]

---

[2] LLE also stands for Loss of Life Expectancy. There are unfounded rumours about the sad story of Lenny Lewis-Engelman being a hoax invented by the author to illustrate a probabilistic concept.

[3] (Cohen 1990) offers more precise data that are a bit less depressing. A pack a day means a LLE of 8.6 years for men and 4.6 years for women. The rest of the figures related to LLE are taken from this excellent text.

After discovering how risky it was, Lenny Lewis-Engelman quit smoking, but wasn't able to persuade his fiancée Helen to do likewise. Our hero made his calculations and concluded that living with a smoker would cost him 50 days of life expectancy, so he decided to break up with his love, however, heartbreaking the decision.

Taking advantage of the extra free time at his disposal after his breakup, Lewis-Engelman took to studying other risk factors. He found out that our excess weight shortens our life: each extra kilogram represents around two months of LLE. So he immediately started to diet, especially when he discovered that every 100 kcal on top of his energy needs (an apple, two low-fat yogurt) deducted 15 min of life expectancy.

Studying further, Lenny was shocked to discover that the LLE related to cardiovascular disease is 5.8 years, so he gave up his job and stayed at home all day to get rid of any stress. LLE related to cancer and heart attack is 4.7 years, and in order to reduce this, he moved to the countryside, where he subsisted on spring water (boiled and filtered) and macrobiotic onions.

In his country retirement, fully devoted to his studies, he discovered, to his dismay, that the loss of life expectancy for remaining single was still higher (6 years) than that of dying from a heart attack. And he had split up with Helen for just 50 miserly days! He phoned her and tried to make up again, but alas, it was too late. The girl had gotten over her disappointment by taking up free climbing and car racing. Lewis-Engelman solved the problem by marrying a girl from the neighboring village. It was a marriage of convenience. The girl came from a poor, uneducated family, which, according to his calculations reduced her life expectancy around 4 years; but on the other hand, a man's LLE compared with a woman's amounts to more or less this period, so he congratulated himself: he had found a perfect compromise.

One day, Lewis-Engelman was returning home after one of his cautious country walks when a violent storm broke out. Our hero didn't worry, knowing that the probability of being struck by lightning is ridiculously low (lower than one in a million). Unfortunately, he didn't anticipate slipping on the wet floor and the fateful fall that caused his death. There's statistics, and there's bad luck.

The Risk of Living Next to a Nuclear Power Station

As a summary of the previous considerations, the table below offers a detailed breakdown of some of the risks involved in being alive. Following Lewis-Engelman's example, let's now calculate the risk of developing a cancer caused by living next door to a nuclear plant.

Loss of life expectancy due to various activities. *Source* (Cohen 1990)

Activity or risk	LLE (in days)
Poverty	3,500
Cigarettes	2,300
Cardiovascular disease	2,100
Being single	2,000
Coal miner	1,100
Cancer	980
15 kg overweight	900
Accidents (combined)	400
Alcohol	230
Road (USA)	180
Flu	130
Suicide	95
Homicide	90
Pollution	80
Occupational accident	74
Drowning	40
Falling	39
Coffee (2 cups a day)	26
Butter with your breakfast	1.1
Hurricanes	1
Plane crash	1
Dam break	1

Let's see. For every extra 30 μSv we receive (more or less the average excess from a nuclear plant, and lower, by the way, than the excess in the neighborhood of a coal-fired plant), our chances that we will contract a fatal cancer increase by 3 parts in 4 million (UNSCEAR 2000). This risk is equivalent to losing 2 min of life expectancy.

To get an idea: for every 10 extra μSv we receive, the chances that we will contract a fatal cancer increase by one part in 4 millions (UNSCEAR 2000). This risk is equivalent to losing 0.04 days of life expectancy (Cohen 1990).

In view of these figures, you might say that educational campaigns encouraging smokers to quit, or promoting a healthy diet, as well as dating agencies, are more beneficial socially than anti nuclear campaigns.

Accidents

Nuclear energy worries the public because of the possibility of an accident that might release a great amount of radioactive materials into the atmosphere, a possibility that the opponents of nuclear power keep going on about. This concern is complicated by the lack of general knowledge about the effects of radioactivity on the human body.

As has already been said, the core of a reactor cannot explode like an atomic bomb. The possible accidents are these: a runaway chain reaction that eventually leads to a chemical explosion (like in the Chernobyl reactor) or a sudden loss of coolant (water, in case of an LWR), which also causes a quick increase of the temperature inside the core (in this case due to the decay of radioactive nuclei, continuing after the chain reaction has come to a halt). This is what happened at Three Mile Island, in Pennsylvania, in 1979, and these are the two most important accidents in history. Chernobyl cost the lives of many people, and is the main reason, together with nuclear tests and the memory of Hiroshima, why nuclear energy is widely rejected. The Pennsylvania accident ended without casualties or releases of radioactivity. In fact, it was a decisive proof of the fact that if they are correctly designed, nuclear reactors are capable of withstanding a situation as extreme as a partial core melt.

Let's review each case in detail.

The Chernobyl Reactor

In contrast with water, graphite (essentially, pure carbon) is such an efficient moderator that it allows the use of natural uranium as a fuel, making enrichment unnecessary and thus saving costs. Graphite was in fact used the first time a chain reaction took place, in the world's first reactor, built by Enrico Fermi under the football field of the University of Chicago in 1942.

On the other hand, a reactor using water both as a moderator and as a coolant has an inherently safe design. The first consequence of a chain reaction is a runaway temperature rise in the core, so the water surrounding the fuel rods evaporates. With the water evaporating, the amount of moderator decreases and the ^{238}U takes care of removing the excess neutrons, even if the control rods haven't been lowered. Therefore the chain reaction stops without human intervention, and the system is robust, unaffected by operating mistakes. The chain reaction is also halted in case a pipe breaks or one of the valves of the primary high-pressure system bursts; this leads to a loss of coolant but also of moderator, without which fission can't go on.

The Chernobyl reactor, however, employed graphite as a moderator and water as a coolant. In this case, both if the chain reaction runs out of control and if there is a loss of water, what happens is just the opposite of what happens in a LWR, because water—in comparison with graphite—acts as a neutron absorber; when it disappears, the chain reaction accelerates, spurred by the efficient moderator, instead of stopping.

So why did the Soviets decide to build a reactor like Chernobyl, given the advantages of the LWR? The answer isn't nice. It was a dual-purpose design, aimed at generating energy... and plutonium for the nuclear warheads in the

Russian missiles. They work with natural uranium (that is, 99.3% of ^{238}U), so the rate of neutron capture by the ^{238}U is much higher than in a reactor with enriched uranium (where ^{238}U competes with ^{235}U), and hence more plutonium is produced.

In a LWR the fuel is contained in a pressure vessel, and refueling is a task that takes about a month and is done at most once a year. But the pure plutonium that is needed for atomic bombs has to be taken out of the reactor a few weeks after being produced, to avoid its contamination with ^{240}Pu and other trans-uranians. So access to the core must be easy. In the Chernobyl reactor, each of the fuel rods was stuck in a tube that could be opened from outside the core, allowing to retrieve the fuel as soon as the plutonium had been formed, without stopping the chain reaction.

However, to make refueling easy and flexible, *the core wasn't protected by the pressure vessel and the concrete bunker which houses an LWR*, but by a containment building, which was not designed to withstand a major accident. In fact, many analyses (Cohen 1990) have shown that the safety barriers of an LWR would have withstood the Chernobyl accident, without leakage of radiation into the atmosphere.

To sum up, contrary to what nuclear opponents keep reciting, an accident like Chernobyl *cannot happen* in LWRs when they are used for strictly peaceful purposes and designed to be safe (and not to produce plutonium). This is due, in the first place, to the self-regulating capacity of water, which allows to extinguish the reaction—without human intervention—when the moderator disappears. In the second place, to the successive containment barriers included in the design.

The Chernobyl Accident

In April 1986, an electrical engineering experiment was scheduled at the Chernobyl plant to improve the performance of the steam turbine. The experiment did not involve the reactor, whose only role was to supply steam to the turbine (in fact, the power needed was below normal), so there weren't any experts in nuclear physics in the team. Operations should have begun at one o'clock in the afternoon on 25th April, but there was an emergency which delayed them until 11 o'clock in the evening. The first step was reducing the output of the reactor to the low level desired for the test (25% of its nominal power). To this end, a certain number of control rods had to be lowered which would absorb the precise number of neutrons for the output to be reduced to 25%. But the delay had caused some stress and an excess number of rods were lowered; the output was reduced to 6% of nominal level, which was not acceptable for the test.

The right thing to do—the obligatory course of action, according to the strict norms prevailing in the West—would have been to cancel the experiment and to rise the reactor power little by little, gradually extracting the control rods during several hours. Instead, the operators tried to force an increase in power by lifting too many control rods in a short time. All rules had been broken, and yet the supervisors decided to carry on with the experiment at one in the morning. As part of the test plan, a series of pumps were activated which increased the flow of water in the reactor. But in a graphite reactor, water acts as an absorbent, so the reactor power decreased still further. You might think that the actors of that ill-fated night were playing a Greek tragedy where the gods control the humans' will to doom them: the operators' reaction was the worst possible. Instead of discontinuing the experiment even more control rods were removed.

At 1:22 a.m. the extra flow of water stopped, but the control rods weren't reinserted. Half a minute later, one of the monitoring computers printed a warning stating the need to shut down the reactor, which was ignored by the operator. At 1:23 a.m., the experiment started, but at the same time the water in the reactor was evaporating and the chain reaction accelerated. Control rods were lowered automatically, but it was already too late. The water started boiling, so the chain reaction speeded up further, building up more high-pressure steam and rising the temperature. There was no time to reinsert the manual control rods (which should have never been taken out). The chain reaction ran out of control and, in contrast to what would have happened in an LWR, there was no way to stop it; the heat inside the core reached a temperature which was 100 times the maximum allowed value. This resulted in the disaster we all know. The explosions were chemical blasts, probably caused by the explosion of hydrogen gas that was formed when the water reacted with the metals in the reactor.

Around 1:30 a.m., fire brigades from Pripyat and Chernobyl arrived at the plant, and at 4 o'clock most fires had been extinguished. Many of the fire fighters received extremely high radiation doses, among other things due to inadequate equipment.

Besides, the graphite acting as moderator in the core had caught fire (graphite burns like coal), so its ashes were scattered into the atmosphere. To extinguish this fire, helicopters were flown that buried the core under tons of boron, sand, mud and lead. The pilots of these helicopters also suffered from lethal doses of radiation.

All in all, according to UNSCEAR, 30 people died (2 in the act and 28 later from the effects of radiation), and more than 100 were injured.

The consequences for the civil population were as follows (UNSCEAR 2000): 116,000 people had to be evacuated from the surroundings of the reactor; all the region suffered from serious economic, social and psychological damage; around 6,000 children and adolescents exposed to the radioactivity released during the accident contracted thyroid cancer (most of them were treated successfully by excision of the thyroid); according to UNSCEAR, there is no scientific evidence of other diseases due to radiation; specifically, no increase in the frequency or malignity of cancer has been recorded. Apparently, the risk of leukemia hasn't risen, one of the greatest concerns for the first years after the accident. An increase of cancer or leukemia cases has not even been recorded among the crews of workers in charge of sealing and cleaning the reactor. UNSCEAR concludes that the vast majority of the population is not likely to experience serious health consequences as a result of the accident.

The United Nations Scientific Committee on the Effects of Atomic Radiation (UNSCEAR) has spent 15 years studying the effects of the Chernobyl tragedy.

Their report, which is available online,[4] is possibly the most authoritative source of information we have. Their conclusions cannot be shrugged off. A lot of people suffered from an accident that shouldn't have happened. But it's not less true that the number of direct victims was much lower that usually assumed—when I pose the question, I often get answers ranging between 1,000 and a million casualties— and that the incidence of cancer, leukemia and abnormal development doesn't seem to have risen among the exposed population.

A more recent and wholly independent study by the World Health Organization (WHO) concludes (Chap. 7) (WHO 2006):

- Among the 134 emergency workers involved in the immediate mitigation of the Chernobyl accident, severely exposed workers and firemen during the first days, 28 persons died in 1986 due to ARS, and 19 more persons died in 1987–2004 from different causes. Among the general population affected by the Chernobyl radioactive fallout, the much lower exposures meant that ARS cases did not occur.
- According to the data, total mortality among Russian emergency workers doesn't differ in a statistically significant way [...] from the mortality of the whole Russian population. However, increases in morbidity and mortality in emergency workers caused by leukemia, solid cancers and circulatory diseases were recently detected.
- Epidemiological studies of residents of areas contaminated with radionuclides in Belarus, Russia and Ukraine performed since 1986, so far have not revealed any strong evidence for radiation-induced increase in general population mortality, and in particular, for fatalities caused by leukemia, solid cancers (other than thyroid), and non-cancer diseases.
- From more than 4,000 thyroid cancers in children and adolescents (0–18 years) diagnosed in 1992–2002 in Belarus, Russia and Ukraine, less than 1% of patients have died from this disease, and the rest were treated successfully.
- Because of the uncertainty of epidemiological model parameters, predictions of future mortality or morbidity [...] should be made with special caution. Significant non-radiation related reduction in the average lifespan in the three countries over the past 15 years remains a significant impediment to detecting any effect of radiation on both general and cancer morbidity and mortality.

Chernobyl was an unmitigated tragedy, but after more than 20 years has passed, we have to revise the statistics and reflect about the casualties related to oil, natural gas, carbon mines or aviation accidents. I don't want to play down the disaster, just put it into perspective. It was a catastrophe that couldn't have happened in any Western country, and which was due to a combination of irresponsibility and bad luck.

---

[4] http://www.unscear.org/unscear/en/chernobyl.html.

The Dreaded China Syndrome

"The China Syndrome" is the title of a 1979 film in which an accident causes the meltdown of a reactor core whose temperature rises up to the point of burning through the containment vessel and then through the body of the Earth until reaching the antipodes, China. The movie would have been just one more in a long list of disaster films—we've had filmed versions of atomic wars, asteroid collisions, earthquakes, volcano eruptions, tsunamis, all kinds of deadly viruses, all sorts of shipwrecks, invasions by uncountable species of aliens—had it not been for the second most important accident in the history of nuclear energy, which happened in 1979, 12 days after the release of the movie: the partial core meltdown of the reactor on TMI, Three Mile Island.

A catastrophe like this is perceived by the public as something almost as deadly as an H-bomb exploding in a city centre. And yet, serious as it was, the accident caused no victims, nor noticeable damage to the environment. Plenty of studies have concluded that the situation at TMI wasn't an exception, but the norm. The reactor was designed to withstand that risk (Cohen 1990).

The reason why the TMI reactor withstood a partial core meltdown was its robust containment building. With more than one meter thickness and a dense mesh of steel beams reinforcing the structure, the building is capable of holding back any leakage, even in case of core meltdown. Besides, it protects the reactor against forces from the outside, such as tornadoes, an aircraft colliding or an explosive charge.

The Chernobyl reactor lacked a building of this kind. Things would have been very different if a similar building had been in place. There would have been no leakage to the atmosphere, no casualties, no irrational fear for years.

What Caused the TMI Accident?

In an LWR, the chain reaction can't get out of control because of the moderating effect of water. On the other hand, even when the chain reaction stops, the reactor core keeps a high temperature due to the decay of radioactive products inside. In normal conditions, the water carries away the excess heat. If there's loss of water, or an LOCA (Loss of Coolant Accident), the core keeps heating up until eventually the fuel bars melt and the radioactivity which is sealed inside the fuel cells is released into the reactor. In order for radioactivity to escape outside, three containment layers have to be pierced: the pressurized vessel, made of 20 cm steel, the bunker made of one meter reinforced concrete, and the outer building, also made of concrete. This has never happened in history.

The TMI accident was due to the combination of two unlikely events: a valve didn't close correctly, and the operators misunderstood the warning issued by the control instruments. This accident, which happened 30 years ago, has led to an immense improvement of the control systems. In recent years, computer technology has been enhanced, and operating protocols are much stricter. The accident has not happened again in the last 13,800 years.[5]

In any case, the core of the Three Mile Island reactor didn't melt completely, as the operators managed to solve the problem in time. It was a near miss, hundreds of which happen in civil aviation—and most of us could tell about one at the steering wheel of our car. And it was a near miss without victims or radioactive leaks, which nevertheless, like anything nuclear, caused panic among the public and was used for decades to "prove" reactors are unsafe.

Black Holes

Operation of nuclear plants is based on safety protocols including redundant emergency systems, highly trained personnel and the accumulated enhancement of each component and each activity, thanks to 60 years experience. In both cases, philosophy is prevention. A lot of talent and money is invested in imagining *what can go wrong* and planning actions to prevent a problem which, very often, has never happened.

These safety measures include resistance to earthquakes, terrorist attacks and thousands of potentially dangerous situations. If we multiply the hundreds of reactors that have been operating for six decades by the number of operation years, the number of reactor years is more 15,000. In this period, there has been *one* catastrophe (in a reactor which was unsafe because of military applications) and a handful of relatively serious accidents which, however, haven't caused victims or noticeable radioactivity leaks.

This doesn't mean that, in case there really is a grade 9 earthquake with its epicenter right under the nuclear plant, the bunker is guaranteed not to break; nor can we be sure that it won't burst if there is a sufficient number of missiles hitting it head on. When these scenarios are discussed, the emphasis is usually on the thousands or millions of potential victims, without taking into account the likelihood of the scenario. This lack of perspective gives rise to ample confusion.

I don't want to imply that people lack the intuition to judge the importance of a catastrophe. A recent example is the tsunami that struck the Indian Ocean in the year 2004, causing more than 230,000 casualties. The number is staggering, but in a way we're able to accept the tragedy without getting obsessed by the possibility of a similar cataclysm happening again, maybe because the inevitability—and also the low frequency—of natural disasters is in our collective unconscious. If we divide the number of casualties in the catastrophe of the Indian Ocean into the population of the Earth, we get a probability of three parts in 100,000, much lower, by the way, of the probability that we'll die due to the impact of the next asteroid.

---

[5] That is, 460 reactors operating for 30 years.

But then, there are other kinds of risks, most of them related to technology—on which we depend in everything, but in which we trust less and less as we perceive it as magic—or to terrorism (perhaps because, like technology, we consider ourselves incapable of understanding or controlling it) that we get much more nervous about. For example, an elementary probability calculation reveals that most safety measures we have to go through at airports are useless and in fact scare the passengers, causing an increased risk perception, rather than reassuring them. Another notable example is the recent commotion stirred up by the operation of the LHC, the great proton collider at CERN. The chances that this research machine might create a black hole capable of swallowing the Earth are about one in 10^{-40}, a figure which is ridiculously small compared with the chance of the most worrying external catastrophe for our civilization (the impact of a large asteroid), and certainly irrelevant compared with the very actual threat of a great disaster linked to climate change.

And yet, the media uproar and the collective neurosis that preceded the start-up of the machine were astounding. The argument put forward then, and which is still being used by the fans of panic holds that, while the catastrophe is certainly extremely unlikely, the number of casualties it would produce (all of mankind, present and future) is so high that both figures balance out. How could you assess a risk whose probability is essentially zero, with a cost that is practically infinite? We don't know. Faced with the imponderable, there are people who propose to solve the problem the hard way, banning the LHC. The same arguments could be used to ban medical research in almost all its branches, including immunology, genetics, neurobiology etc. You would also have to ban nanotechnology, space exploration, radio communication (to prevent an alien civilization from detecting and extinguishing us) and computing in general (given the risk of an artificial intelligence emerging and enslaving us, another science fiction classic).

Most members of the public may be a bit frightened at the beginning—especially thanks to the power of the yellow press—but they stop worrying after a while, and rightly so. You don't need to be an expert in statistics to be able to separate the grain of truth from the chaff of imaginary catastrophes.

Nonetheless, a great many of the arguments held against nuclear energy are as absurd as the fear mongering over the black hole at CERN. But in everything nuclear, the common man is easily influenced. You read apocalyptic reports about Chernobyl and end up believing that the episode cost more lives, directly or indirectly, than the tsunami in the Indian Ocean. You read the never-ending accusations, mostly unfounded, made by activist groups, and develop the notion that nuclear plants are unsafe and dangerous. Once and again, we see photographs of the ominous barrels, with a yellow trefoil signaling their radioactive content, and we imagine radioactivity as a kind of lethal fluid that spreads through the air like a virus and poisons us.

I don't pretend that radioactivity is harmless. A krypton leakage from the core of a nuclear reactor can have pernicious effects on the people in the neighborhood—remember that any leakage is rapidly diluted in the atmosphere, and this is the reason why there are practically no traces of the great nuclear tests performed during the Cold War—so reactors are designed to prevent these leaks, taking as much care as aviation engineers take in designing planes whose wings don't break. Aviation is safe, not because it's impossible that we'll suffer a mortal accident, but because the odds are so low (on average, we have to take a flight every day for around 21,000 years to get the winning ticket) that the risk doesn't worry us, except when one of these rare accidents happen and the press keeps terrifying us for weeks. Demanding a ban on nuclear energy based on such abysmally low probabilities is as absurd as demanding a ban on commercial aviation for safety reasons.

I find it difficult to understand the furiously anti nuclear attitudes of Greenpeace and other like-minded groups who stubbornly insist on remote or unlikely accidents and get obsessed with minimal probabilities, disregarding the fact that we are experiencing a serious threats for all of mankind: climate change. Their attitude is not too different from that of a shipwrecked person dying of thirst who rejects the water springing from a well he has miraculously come upon for fear the excess of calcium might harm him.

Radioactive Waste, an Intractable Problem?

The same as many other problems related to nuclear energy, the issue of radioactive waste has a heavy symbolic and semantic charge. We all know the words: waste, radioactive, high activity. We are all familiar with the tri-blade symbol. We all have heard they last for millions of years. We all know it's *extremely dangerous*.

It's certainly dangerous if you decide to have a bath in the pool where it's stored after being retrieved from the reactor, or prepare a snack with it. There's nothing unusual about the waste of an industrial activity being harmful or highly toxic (in fact it's almost the norm); it's just that we need a guarantee that this waste is kept under strict control. Some examples of toxic substances handled by the industry on a day-today basis which are fundamental for our society are: hydrogen chloride and hydrochloric acid (cleaning, treatment of metals and galvanizing, refining and manufacture of a great variety of products); cyanide (extremely poisonous chemical substance, industrially used for electroplating zinc, gold, copper and silver); ammonia (to obtain fertilizers, textiles, plastics, explosives, pulp and paper, food and drink, household cleaning products, coolants and other products) and arsenic (insect killer, weed killer, wood preservation, another powerful poison).

A coal-fired power plant releases 11 tons of CO_2 per minute into the atmosphere, emits sulfur dioxide (acid rain), nitrogen oxide and ashes, and besides traces of all kinds of metals, like lead and cadmium. Last not least, the ashes that are emitted contain uranium, radium and thorium, present in a few parts per million in coal (of which, as we have seen, billions of tons are burnt), and are a source of radon, a radioactive gas that is the most dangerous *natural* source of radioactivity. In contrast, a 1,000 MW nuclear plant releases nothing but steam into the atmosphere. The waste from a coal power plant weighs five million times as much as the waste from a nuclear plant.

Temporary Storage

We have already seen that the spent fuel is kept in the pools of the nuclear plants for 10–30 years, where its radioactivity diminishes by a factor of at least 100. One of the exaggerated complaints that is often voiced is the "serious problem" of waste being stored in the power plants. In fact the only problem is space, because the storage pools eventually fill up and part of the fuel has to be taken out. But the longer it remains in the pools, the lower its radioactivity, and the easier it is to handle. Keeping the waste in temporary storage also means that it may be possible to reprocess it at some moment.

Temporary waste stores are designed to withstand all kind of disasters, including floods, tornadoes, missiles and extreme temperature changes. When the waste is put in dry barrels after ten or more years in the pool, the double layer shielding of the container stops gamma radiation completely; only heat (equivalent to household heating) is released to the outside.

By the way, James Lovelock has publicly offered the garden of his house to store a few dry barrels or some vitrified waste, with a view to making use of the thermal energy they provide. This idea, though it has deliberately been expressed in provocative terms, is no nonsense. It would be perfectly viable to use the heat given off by radioactive waste in a cogeneration system for urban heating. In fact, the idea has been patented,[6] and the only reason for not putting it into practice is political, to express it somehow.

High Radioactivity Waste

In Chap. 9 (Figs. 9.4 and 9.6) we saw that both the activity and the radio toxicity of high radioactivity waste (above all, ^{137}Cs and ^{90}Sr) don't fall below natural uranium until several hundred years have passed. From then on, trans-uranians dominate the activity, which needs another 10,000 years to reach natural uranium levels. If future fast neutron reactors consume the trans-uranians, the problem of storing the waste is measured in centuries. Otherwise, it's millennia.

Where can we store this in any case? Bernard Cohen (1990) and James Lovelock (2000) suggest two quite radical alternatives to get rid of it. The former recommends dumping it into the deep sea. The latter, spreading it in the Amazon

---

[6] http://www.freepatentsonline.com/6183243.html.

forest and other natural environments he considers sacred, in order to protect them from human invasion.

Cohen's proposal is absolutely sensible from a scientific point of view, though it may sound like blasphemy. Vitrified waste is housed in stainless steel containers that can last for thousands or tens of thousands of years without corroding, and glass, in any case, doesn't dissolve in water. All radioactivity waste produced up to now by the whole nuclear industry fits more or less into about 100 hectares. Assuming that the number of nuclear plants is multiplied by ten, and that they will continue operating for another five hundred years, we would fill about 10 square kilometers, the sea floor occupying more than 350 million square kilometers. If the containers were thrown randomly into the Pacific Ocean, coming across them would be more difficult than finding a needle in a haystack. I can conceive of no reason for not solving the "intractable waste problem" in this safe, simple, economical way.[7]

Is It Dangerous to Bury Radioactive Waste?

There is a less radical—or more politically correct—alternative, storing the waste in an appropriate geological repository. This choice, by the way, was the one favored by Nature when she concluded her fission energy experiment—after operating for 2 million years—and it has worked flawlessly for 1998 million years. There are plenty of possible repositories on Earth, all of them spacious, dry and stable enough. Finding them isn't a technical problem: it's, once again, a political problem.

Yucca Mountain (Fig. 10.3 left) is a relevant example. This is the site proposed by the US Department of Energy (DOE) as a geological repository. It's a mountainous region in Nevada, about 100 km north east of Las Vegas, featuring a dry climate, removed from population centers, with limited access, stable geology, a deep aquatic bed and no known surface streams. The mountain has been studied by hundreds of geophysicists for the last 20 years, so it's probably one of the best understood places geologically. Some of the peculiarities that have been well established are low rainfall, slow movement of water in the mountain rocks (1 cm per year) and a stable geology, evidenced by floating rocks (Fig. 10.3 right), which suggest there has been low seismicity for many thousands of years. The studies performed during two decades prove that the risk of volcano eruption, erosion or other geological processes is minimal.

---

[7] But then José Díaz, professor of atomic and nuclear physics at the University of Valencia, suggests a weighty reason. Much of the radioactive waste will be useful in the near future, either as a source of fission energy—the case of plutonium—or in technological applications—Americium-241 would be an example of this. So he concludes that they should rather be stored in geological deposits, which he considers genuine mines of valuable raw materials and not ancient cemeteries or radioactive garbage dumps.

Fig. 10.3 *Left* Yucca Mountain geological repository, being studied by the DOE since 1987. *Right* Floating rocks on the mountain, an evidence of lack of geological movements. *Source* (DOE 2008)

The repository will be placed at around 300 m underground and 300 m above the aquatic bed, in such a way that it is protected both from earthquakes and from underground streams (and possible terrorist attacks). Besides, the mountain lies in a wide basin, surrounded by highlands that prevent the underground streams flowing under the mountain—much deeper than the repository—from discharging into large watercourses or into the ocean.

To top it off, access to the site is restricted, controlled by the army. The air space is closed and monitored by the Air Force. The nearest town is twenty-two km away. There are few places on Earth as calm, remote and protected.

In this context, it may well be worth noting that, if we took similar precautions with other activities related to energy, the world would suffer a global collapse. We would have to close down coal mines (they are demonstrably dangerous), oil wells (an obvious terrorist target), natural gas regasification plants (liable to explode) and would have to halt the expansion of wind energy waiting for several decades of studies to prove that the various harmful effects attributed to wind turbines (interference of electromagnetic waves, landscape degradation and danger for birds, among others) are acceptable.

If the last item raises some hackles (to halt such a clean energy just for aesthetical reasons or because it kills a bird once in a while?), consider that radical anti nuclear campaigners claim the 20 years of studies in which a battalion of geologists have taken part, plus the safety certification for the next 10,000 years, are insufficient. The habitual argument resorts to an impossible water current entering the repository, dissolving the waste, dragging it to an aquifer outside the valley and poisoning the people.

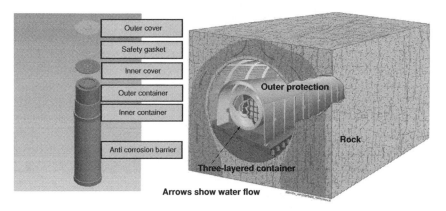

Fig. 10.4 Nuclear waste encapsulation in Yucca Mountain. *Source* (DOE 2008)

Figure 10.4 shows how radioactive waste is encapsulated. The inner layer is the vitrified waste. The next layer is a metal container, usually made of stainless steel or a titanium alloy, which is very corrosion resistant. Then there's a layer of stabilizer and two isolating layers. Finally, all spaces between the cylinder and the rock are filled with some absorbent material that will catch any leakage from the cylinder.

To sum up, there are five safety barriers:

1. Waste is buried in a region whose safety has been certified by exhaustive geological studies which, as we have seen, ensure against all kind of risks. Seismic, volcanic and water stability, including the absence of underground streams.
2. If geologists have made a mistake and an unforeseen water current appears, the rock itself offers a huge protection, because water has to dissolve it in order to reach the waste embedded in it.
3. In case the water does reach the chambers where the waste is stored, it encounters the filling material between the rock and the cylinder, usually clay, which expands sealing the cylinder and isolating it from humidity.
4. If the water manages to penetrate through the clay, it finds the titanium (or steel) barrel, highly resistant to corrosion. The design of the barrel, which has passed strict laboratory tests, guarantees it will resist corrosion for thousands of years (and once this time has passed, 99% of the toxic waste will have decayed). This is the reason why Bernard Cohen wouldn't mind throwing it into the sea.
5. If the water does manage to corrode the barrel, there's the vitrified waste. Glass is extremely difficult to dissolve, we have archaeological evidence (glass artifacts from ancient Babylonia) that it has resisted for thousands of years in a riverbed without dissolving.

And what if nonetheless some waste gets dissolved? That's the reason why Yucca Mountain is in a basin: it means the water can't leave the valley.

If you multiply all the odds: water appearing unexpectedly, dissolving the rock, dissolving the absorbing material, dissolving the titanium alloy, dissolving the glass, dragging the waste and climbing uphill, the result is zero. Well, to be more precise, it isn't exactly zero. Not even the probability that the Earth may disappear engulfed by a black hole, a threat that troubles more people than you might think, is zero. But worrying about such an unlikely event is as absurd as agonizing over cosmic singularity.

An Unsolved Problem?

To sum up, radioactive waste doesn't pose an unsolved problem. It isn't difficult to find adequate geological sites, to condition them properly and to store the waste in the ultra safe way we have described.

On the contrary, there's an unsolved problem with climate change, whose consequences may be devastating for our civilization; or, to put it in different words, there's an unsolved problem with CO_2 emissions due to the annual burning of 6,000 million tons of coal, 3,000 million tons of oil and 4,000 million tons of natural gas. It seems incredible that, in view of a threat as serious as global warming, there are still people with a disposition to worry about imaginary problems. To speculate about the possibility of waste escaping—by magic— through the multiple layers it is contained in, and causing a disaster in one 1,000 or 10,000 years, while draughts, floods and plagues may cause millions of victims in the next century, is childish and unwise. And yet, that's precisely the attitude so many renowned eco-dogmatics stick to.

However, it's an undeniable fact that public opposition to nuclear waste repositories is much stronger than opposition to nuclear power plants—it's certainly the case in France—although, when it comes to safety, an active reactor, however, safe it might be, always entails a greater risk than passive waste shielded by multiple protecting layers and buried underground. If the risk of a reactor releasing a noticeable amount of radioactivity to the atmosphere is tiny as of today, the risk associated to buried waste is zero for all practical purposes.

So how can we explain public opposition? Possibly it's the dreaded NIMBY effect: Not In My Back Yard.

The reason why storing radioactive waste is troublesome has got nothing to do with safety—a stainless steel barrel full of glass embedded in a rock at 300 meters underground poses less of a risk for you than crossing a street—but with the fact that it doesn't produce added value. What I mean is this: a nuclear plant can be perceived as a problem, but also as an opportunity—it creates a lot of jobs and sparks the local economy. By contrast, a waste repository—especially if we think of it as a garbage dump or a cemetery—is perceived as a problem—a restricted area, rumors about unsafety, fear of terrorist attacks—*in exchange for nothing*. The immediate reaction is "not in my backyard". It's visceral, selfish, shortsighted—and absolutely human. And easy to exploit.

An argument that is often given by anti nuclear apostles is the "irresponsibility" of bequeathing radioactive cemeteries to future generations. The concern about humans in 1,000 years—who most probably will have the technology to burn all the waste inherited from us, if they so wish, though it's more likely they will leave them in peace, or follow Lovelock and scatter them in the Amazon forest to scare poachers away—is commendable, but it's more commendable to worry about our children and grandchildren, the ones who will have to face the worst effects of climate change if it isn't solved in time.

Is waste an intractable problem? In a way it is: there's no way to get rid of it, uranium is everywhere. The fact is well known by the scientists carrying out experiments that require very low radioactivity conditions (for example the search for dark matter or for certain properties of neutrinos that might reveal the cosmic asymmetry between matter and anti matter) which are extremely hard to obtain. Everything surrounding us is at least slightly radioactive. This is because our planet is just a giant waste deposit, originating from one of the greatest nuclear accidents in recent times: the supernova explosion that formed the solar system and left such an amount of uranium in the Earth's core that it's still red-hot.

References

Cohen, B. L. (1990). *The nuclear energy option*. New York: Plenum Press.
DOE. (US Department of Energy, 2008). *DOE studies of the Yucca Mountain geological site*. http://www.ocrwm.doe.gov/ymrepository/.
Lovelock, J. E. (2000). *Gaia: A new look at life on earth*. ISBN 0-19-286218-9.
UNSCEAR. (2000). *Report of the united nations scientific committee on the effects of atomic radiation to the general assembly*. http://www.unscear.org/docs/reports/gareport.pdf.
WHO. (2006). *Health effects of the Chernobyl accident and special health care programmes*. http://www.who.int/ionizingradiation/cher-nobyl/WHO%20Report%20on%20Chernobyl%20Health.

Chapter 11
The Anti Nuclear Litany

Litany (from Latin litania)
A usually lengthy recitation or enumeration
Any long or tedious speech or recital

Fifteen Lies

Can nuclear energy become a viable alternative to fossil fuels? Almost all of French electricity is generated in nuclear plants, as is half of the electricity in Ukraine, Sweden and Belgium, a third in Finland and South Korea and a fifth in many other countries, among them Spain, Germany, the UK and the USA. The only renewables as of today capable of supplying a portion of the electric mix that is comparable to nuclear energy are hydroelectricity and, to a lesser degree, wind energy. Waterfalls account for 100% of electricity in Norway, Uruguay or Paraguay, but this resource, as is usually the case with renewables, is unevenly distributed. Some countries have a lot, others few. In Spain, hydropower and wind energy provide 10% each, so together this is similar to the nuclear portion.

In view of these figures, it's not far-fetched to imagine that nuclear energy may become more important in the future and gradually replace fossil fuels, specifically coal. Besides, you have to take into account its potential for hydrogen generation, one of the few alternatives to oil seriously considered as such.

The advantages of nuclear energy are obvious: first of all, the same as renewables, it doesn't release CO_2 into the atmosphere. Unfortunately, organizations like Greenpeace are unable to acknowledge this obvious fact and claim (Greenpeace 2008b):

> The process of nuclear fission does not release carbon dioxide (CO_2), but all the related previous activity does: uranium mining, for instance, requires a great deal of transport and machinery emitting, as a whole, more CO_2 than generating renewable energies.

Figure 11.1 makes things clear. Compared with fossil fuels, all alternative energies, nuclear and renewables, release a ridiculous (if you take the minimum values from different studies mentioned in (Boyle 2003)) or very small (even if you take the maximum, except for hydropower) amount of CO_2. Nuclear energy

J. J. Gómez Cadenas, *The Nuclear Environmentalist*,
DOI: 10.1007/978-88-470-2478-6_11, © Juan José Gómez Cadenas 2012

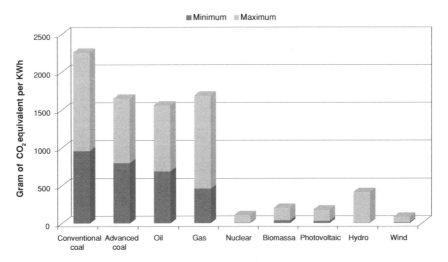

Fig. 11.1 CO_2 emissions to the atmosphere for different energies. *Source* (Boyle 2003)

releases less (indirectly) than photovoltaic, biomass or hydroelectric energy and about as much as wind energy.[1]

A second advantage, which nuclear energy shares with fossil fuels, is reliability and rapid response. The same as a coal or natural gas power plant, a nuclear plant is available when it's needed, and the large Parsons turbines, on which thermal electricity generation is based, can provide more (or less) power, upon demand. Of course, this doesn't happen with renewables such as wind or solar energy; hours of sunlight or wind are limited.

The unavoidable intermittence of the wind results in poor performance of wind energy as compared with nuclear. While nuclear plants perform between 85 and 90% of the time, in the case of wind parks the figure is reduced to 20%, that is, five turbo generators have to be installed to assure that one is working continuously.

The press doesn't tell us about these advantages, instead we usually listen to the tiring anti nuclear litany, recited, among others, by Greenpeace. In a recent article published in the Spanish newspaper El Mundo, with the nice title "the fifteen lies about nuclear energy" they say (Greenpeace 2008a):

1. *Nuclear energy is unsafe:* the ecologist NGO reminds us of the Chernobyl tragedy and radioactivity's destructive power.
2. *Nuclear energy can cause conflicts:* it's a potential terrorist target.
3. *Nuclear energy is dirty* because it's dangerous, as nuclear waste lasts for tens of thousands of years and waste management hasn't been solved yet.

---

[1] For the interested reader, I suggest Chap. 13 of the excellent book by Boyle Everett and Ramage [2003]. Though the authors don't sympathise with nuclear energy, they stick to the facts, something that Greenpeace seems unable to do.

4. *Nuclear energy is expensive* and can only survive in countries where it receives important state subsidies. Greenpeace quotes an MIT report according to which in the current situation "electricity of nuclear origin is not competitive".
5. *Nuclear energy isn't accepted by the public:* according to Greenpeace, the majority of people reject it.
6. *Nuclear energy isn't necessary in order to replace fossil fuels:* the examples given are Germany and Sweden, countries with no nuclear energy, to prove that it isn't necessary for electricity generation and in order to attain the goal of reducing emissions.
7. *Uranium isn't an abundant fuel:* fissionable uranium-235 is running out, what remains can only last for 70 years, given the current number of nuclear plants.
8. *Nuclear energy isn't renewable:* they say "The nuclear lobby have made us believe nuclear fission is a renewable energy, and spent nuclear fuel, which is highly radioactive, can be recycled. Obviously, there's no basis for these claims."
9. *Nuclear energy isn't a complement for renewable energies*: clean energy sources can by themselves meet the demand of a country such as Spain, according to a report commissioned by Greenpeace carried out by the "Universidad Pontificia de Comillas".

To begin with, the reader is stricken by clichés such as "radioactivity's destructive power". What is worse, there are statements that are untrue, in particular those referring to Germany and Sweden as countries free from nuclear energy. In fact, Germany and Sweden are among the world's first nuclear powers: Germany has almost 30% nuclear energy in its mix, Sweden next to 50%. In these countries, the same as Spain, there has been a long moratorium on the development of nuclear energy, motivated by eco activist protests and the availability of cheap, abundant fossil energy (in the case of Germany; in Sweden, the other 50% is hydro energy). Besides, their quote from the prestigious study by the MIT (2003), whose conclusions *favor nuclear energy*, is taken out of context.

Another statement which is wrong: "The nuclear lobby have made us believe nuclear fission is a renewable energy, and spent nuclear fuel, which is highly radioactive, can be recycled. Obviously, there's no basis for these claims." In Chap. 9 we saw the principle on which waste recycling is based: plutonium and other trans-uranians are split by fast neutrons. Sustainability is made possible thanks to the transmutation of uranium into plutonium. The fact that the spent fuel is radioactive is not related to recycling capability. The PUREX process has existed for decades, FNRs have operated successfully in the past (Super phoenix among others) and are scheduled for Generation IV, and in fact the pressure to recycle waste originates in academic and scientific circles rather than in nuclear industry. A crash course in basic nuclear physics is something the author of these grave assertions would certainly profit from.

In the preceding two chapters we have also examined why other accusations are groundless. Among other things, we have revised the Chernobyl accident, its consequences and the reasons why an accident like this can't happen in current plants, least of all in new generation plants. We've also talked about waste, the proliferation of nuclear weapons and terrorists. To complete the analysis of the litany, we still have to refer to the price of nuclear energy, the abundance (or scarcity) of uranium, the acceptance by the public and the possibility that renewable energies are sufficient to meet the demand of a country like Spain or, why not, of all the world. This last point is dealt with in the last chapter. We'll now have a look at the other three.

France and Nuclear Energy

One of the routes I have traveled most often during the past 25 years is the one joining Valencia and Geneva. I started traveling in 1983, with my first summer student grant at CERN, and kept doing so for the ten long years I worked in the European laboratory, and almost every year since my return to Spain.

Heading to Geneva from Valencia, you cross the frontier with France at La Junquera and drive along the French Midi on your way to the Alps. There's an image you often come across in the landscape bordering the six hundred something kilometers of highway: the twin towers with white clouds rising up into the skies. They're nuclear plants, of which there are no fewer than 56 in our neighboring country (compared to seven in Spain).

In contrast with Spain, in France nuclear energy is not just accepted by the population, it is *popular*. I have lived in this country for many years and have met very few people opposing it, but a lot of enthusiasts.

In 1973, at the time of the first oil crisis, the French government assessed its dependency on the OPEC countries and came to the conclusion that the situation was desperate. France, just like Spain, almost completely lacks fossil resources. It has neither gas nor oil, and very little coal. The French reaction to the OPEC crisis was decisive. In the following 15 years, 56 nuclear reactors were built, at a rate of almost four a year.

There was some dissent in the 70s, but the French nuclear program has continued being accepted by society.[2] If you read the articles published by Greenpeace Spain, the only possible conclusion is that our neighbors suffer from some strange disease that has infected the whole country and prevents them from realizing an obvious truth, or either the government is capable of brainwashing 60 million people.

In fact, the reasons are mucho more down-to-earth. On the one hand, there's the French ego, the pride of belonging to a great country and with it the refusal of having to bow to the OPEC. Another factor favoring the acceptance of nuclear energy in France is its long scientific and technological tradition. We're talking about the native country of Lavoisier, Carnot and the Curie couple, a country

---

[2] In exchange for some scapegoat, like the fast neutron reactor Superphoenix, see Chap. 9.

where people traveled on high-speed trains (the famous TGV) at a time when in Spain you might know what time the daily express was going to leave, but the arrival time was uncertain.

Consequently, France is a country where scientists and engineers are highly regarded and where a huge number of civil servants and people holding high positions, administrative and political, have studied science, for example at the renowned *Ecole Polytechnique* in Paris. Compare this to Spain, a country where scientists are conspicuous in the political scene—by their absence.

Then, we have to take into account the efforts French authorities have made in order to explain the benefits of nuclear energy to the population. More than 10% of the French (that is, 6 million) have visited a nuclear power plant. And yet, the polls show that most of the people don't understand nuclear technology well, nor are they capable of assessing the risks related to waste or the possibility of an accident. They just trust their technocrats.

And there's something else. What French people do understand, much better then Spanish people, is that they haven't got any fossil fuels and therefore life would be much harder without nuclear energy. For them, the added benefit is being one of the European countries that contributes least to climate change. If you look at it from this angle, it appears that French nuclear ecologists are doing a good job. They worry about important things.

In conclusion, what's needed for the Spanish population to accept nuclear energy is more information, as well as countering the dogmas spread by eco-fundamentalists—the purpose of this book. It's also a question of common sense. The famous slogan our neighbors use to justify nuclear energy ("no oil, no gas, no coal, no alternative") is applicable to our case. All the more because of the fact that it also entails fighting climate change.

Is Uranium a Renewable Resource?

What I want to deal with next is the alleged scarcity of uranium. To begin with, it's worth noting that uranium is as common as brass or zinc. Like it happens with all metals our society consumes, mining companies and the metal consuming industry carry out estimates, revised annually, of the reserves which are available at a certain price. The "reasonably proven reserves" are estimated at 5.5 million tons[3]. At a consumption rate such as the current one, 65,000 tons a year, you get 77 years.

But what does "reasonably proven reserves" mean exactly? As we saw in the chapters devoted to fossil fuels, the term "reserve" doesn't apply to the total amount of the resource, nor to the known amount, but strictly to the *amount of known resources that can be exploited commercially.*

---

[3] http://www.world-nuclear.org/info/inf75.html.

The correct reading for 5.5 million tons of "reasonably proven *reserves*" is "the amount of uranium that is known and can be extracted at a price below 130$ per kilogram". In fact, most of these reserves are estimated at a price of 80$ per kilogram.

Is uranium expensive at the moment? In 2007, one ton of coal costs around 86$ in the European market, more or less the same as a kilogram of uranium. At first sight, and this is one of the most frequent mistakes made, uranium seems to be one thousand times as expensive as coal!

But we mustn't forget that 1 kg of uranium provides as much energy as 3,000 tons of coal. In other words, in energy terms, uranium is 3,000 times cheaper than the cheapest of fossil fuels.

In fact, the price of electricity depends to a great degree on the investment in construction and infrastructures, plus all the cost associated to the processing of waste. At 100$ a kilo, uranium contributes 2.5% to the total price of electricity. This implies that—in contrast to what happens with natural gas and even with coal—uranium can be paid for at up to 400$ a kilo without its contribution to the price of electricity being more than 10%.

Why is this detail so important? Because reserves only measure the amount of uranium available at a certain price. If the price we are ready to pay goes up, so do the reserves.

Just a moment… we live in a finite world, don't we? After all, there's only so much uranium (or iron, zinc, gold, cobalt, manganese, platinum or copper, to mention just a few metals used industrially). What will happen the day we've consumed it? The price may rise, but what can be done when it has run out?

Another way to formulate this argument is to say that uranium (like iron etc.) isn't renewable, as neither are fossil fuels, but that there are finite deposits bound to run out some day. Is that so? Not really. Fossil fuels stem from the decay of organic matter in very particular conditions (Chaps. 4 and 5), while metals are the raw material our planet is made of, and there are huge amounts of them. To give an extreme example, you just have to dig—dig a few kilometers, admittedly—to get to the Earth core and find enormous deposits of iron, nickel and other minerals, including uranium and thorium.

But there are less radical solutions. Uranium, like all metals, is extracted from mines whose streaks contain a certain proportion of the metal. Canadian mines have streaks with spectacular concentrations, of around 20%, but in general a streak with a 2% concentration is considered high-grade, and currently it is cost-effective to exploit even very low-grade streaks, such as the ones in Namibian mines, with concentrations of 0.01% (this means only a hundred parts per million, (ppm) of the streak is pure uranium). On the other hand, uranium is so abundant that it's found at concentrations of five parts per million (5 ppm) in granite and 3 ppm in common rocks. Uranium can also be found dissolved in seawater, at 300 parts per billon (ppb).

Can uranium be extracted from the sea? Of course it can. It was M. K. Hubbert himself who put forward the idea, proven quantitatively by Bernard Cohen (1983). Several research programs, specifically in Japan (Seko 2003) have extensively shown its feasibility. The idea consists in using a fabric that filters the uranium.

The fabric is confined in large boxes and sunk in the ocean. The Japanese researchers placed their prototype at 20 m depth, 7 km away from the coast, and left it there for 240 days, obtaining one kilogram of "yellow cake".

At what price? Less than 500$ per kg seems reasonable, especially if the technology is applied at large scale; the contribution of uranium to the cost of a nuclear kilowatt-hour would be under 10% *with conventional reactors and an open cycle.* However, we've already seen that the future will probably bring fast neutron reactors, which implies that the uranium yield can be multiplied by a factor of 100 (as not only ^{238}U, but also ^{235}U would be consumed). Combining RNRs and uranium recovered from the sea, you get... *a renewable source,* given that the oceans house 1.4 quintillion (1.4×10^{18}) tons of water, containing about 4,600 (4.6×10^{9}) tons of uranium. Using conventional reactors, the resource would run out in around seven hundred thousand years. With RNRs, there would be enough for seven million years, all this without taking into account that uranium is in fact renewed in the sea, as there is a constant flow of 32 tons a year coming from the rivers. Bernard Cohen (1983) considers that an extraction rate of 6,500 tons a year (ten times the current global consumption) is in fact *sustainable.*

By the way, the figures I'm giving are taken from articles published in peer-reviewed magazines (Cohen 1983; Seko 2003), whose authors, like the author of this book, have no relation whatsoever to the "powerful nuclear lobby".

Besides, exploiting marine uranium is by no means the only possibility. The International Atomic Energy Agency IAEA has published its estimates of conventional resources in its "Red Book". According to them, the conventional resources, at prices below 130$ a kilo (that is, the "reasonably proven reserves" plus the likely discoveries of streaks that can be exploited profitably) amount to between 16 and 20 million tons (IAEA 2001; Price and Blaise 2003), which would push the availability of *cheap* (in fact dirt cheap) uranium to between 300 and 500 years (or tens of thousands of years if it's recycled, using RNRs).

This means the concept that the planet's resources are finite and will be running out soon, which was so fashionable during the 60s, is much less obvious than it appears. Undeniably, there is a physical limit for metals, as the case of uranium illustrates, but this limit recedes continuously as our technology improves.

A famous anecdote is right to the point. In 1980, Paul Ehrlich, a celebrated ecologist, and Julian Simon, an also very well known economist, entered a wager, which would bear their names: the Simon-Ehrlich wager. Ehrlich was persuaded that the Earth's resources were on the brink of depletion, and predicted that the price of commodities would increase inexorably. Simon, on the other hand, argued that just the opposite would happen; he was persuaded that the commodities Ehrlich had in mind weren't scarce but plentiful. Their diverging opinions led to a wager centered on the price of five metals: copper, chromium, nickel, tin and tungsten. Simon would pay if the inflation-adjusted price of any of these metals had risen in 10 years. Ehrlich boasted, "The lure of easy money is irresistible". In 1990, the price of each of the five metals had come down, and the ecologist had to pay.

Another similar example. In 1970, the known reserves of copper were expected to last for just 30 years at the production rate of the time, which made many analysts wonder if the resource would be able to meet the demands of the telecommunications industry by 2000. But in 1994 the production of copper had doubled, and there was still enough for another 30 years.

I can't conclude this section without offering a similar bet to the enthusiastic Greenpeace activists. I bet 100 Euros that in 10 years, the production or uranium will have increased—as there will be more nuclear stations operating—and the reasonably proven reserves will remain constant or will have grown compared to today.

The Price of Nuclear Energy

How much does nuclear energy cost? This is not an easy question to answer. If we mean the *current* price per kilowatt-hour, the answer is that both in Spain and in France nuclear electricity is competitive with electricity from fossil fuels (3.5 c€/kWh in Spain, 2.5 c€/kWh in France compared with free market prices of 4.7 c€/kWh and a consumer price of around 0.1 c€/kWh), and much cheaper than electricity from renewable sources (Chap. 12).

On the other hand, both Spanish and French nuclear power plants were built quite a long time ago (between 1971 and 1988 in Spain) and therefore have had several decades to write off their costs. Hence it makes sense to wonder about the price of a newly built nuclear plant and how it compares to other primary energy sources.

The result of one of these studies, published by ENDESA, the main electricity company in Spain, can be seen in Fig. 11.2, comparing the construction of super critical and ultra super critical coal-fired power plants (where water is kept at a pressure and temperature that blurs the distinction between liquid state and steam, giving rise to great thermal efficiency), natural gas, nuclear, hydroelectric, wind powered and photovoltaic plants, under two assumptions: either, gas being cheaper than coal, which is called gas priority (this can happen if the taxes on CO_2 emissions increase sufficiently), or coal being cheaper, that is, coal priority (which seems rather likely given the high prices of natural gas). As can be seen, according to this study nuclear energy is still the cheapest, while photovoltaic energy is almost 100 times more expensive. A more recent study was carried out in 2008 by the Spanish economy professor Santos Ruesga (2008). His cost estimate is extraordinarily detailed, and rises to 3 billion euro per power plant, thus increasing the price by 50%; but even so, as you can see in Fig. 11.2, nuclear energy is still competitive with energy obtained from fossil fuels, and considerably cheaper than renewables.

In its rebuke, Greenpeace quotes the prestigious MIT study "The future of nuclear energy" (MIT 2003), as stating, "nuclear energy isn't competitive".

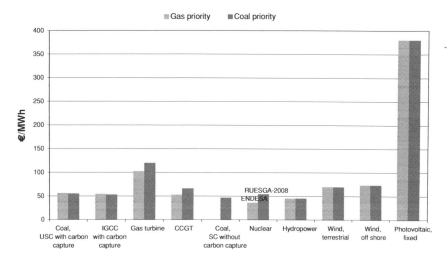

Fig. 11.2 Cost of electricity generation by technology. Source (Nuclear 2008) and (Santos Ruesga 2008)

The MIT study is available on line, so the interested readers can check the contents for themselves. Electricity prices range between 4.2 and 6.7 $c/kWh. (depending on optimization of construction prices). The price of electricity from coal is 4.2 c$/kWh (without taxes on CO_2 emissions) and can reach 9 c$/kWh with a tax of 200$ per ton (if the current price for CO_2, which is around 100$ per ton in the EU, is applied to emissions from thermal plants, the kWh would cost about 6 cents). For natural gas, the scenarios range between 5.6 and 7 c$/kWh, depending on the tax (based on the foreseeable high prices for natural gas).

We see that the statement "nuclear energy isn't competitive" is taken out of context. To begin with, if a great number of nuclear plants are built and techniques are standardized (as the French did 30 years ago), the price of the nuclear kWh goes down and becomes competitive with coal *without tax* and cheaper than natural gas (whose price has kept going up since the MIT study). Besides, any reasonable scenario envisaging a plan to fight climate change will include taxes on CO_2 emissions, which in fact raises the price of the fossil kWh as compared to nuclear.

To sum up, that's one more claim in the litany that turns out to be false. There are others I haven't bothered taking down, like the claim that nuclear energy doesn't generate jobs (Santos Ruesga reckons 200,000 jobs a year for 20 year in his project (Ruesga 2008), or that it isn't competitive for emerging economies (India and China are embarking on powerful nuclear programs).

Some thoughts to conclude this chapter. You can argue about the pros and cons of nuclear energy, or of anything else, but you must always resort to reasonable and well reasoned arguments, and above all, be honest. In contrast, the exaggerated, inaccurate or false claims the litany has been showering on us for decades shows bad faith or ignorance.

References

Boyle, G. (2003). *Energy systems and sustainability*. New York: Oxford Press.
Cohen, B. L. (1983). *Breeder reactors. A renewable energy source*. http://www.sustainable nuclear.org/PADs/pad11983cohen.pdf.
Foro Nuclear. (2008). http://www.foronuclear.org.
Greenpeace. (2008a). *Informes renovables 100%*. http://www.greenpeace.org/espana/reports/informes-renovables-100.
Greenpeace. (2008b). Las quince mentiras de la industria nuclear. http://www.elmundo.es/elmundo/2008/11/11/ciencia/1226403255.html.
IAEA. (2001). Uranium 2001: Resources, production, and demand, Paris, France: OECD Nuclear Energy Agency and International Atomic Energy Agency, 2002.
MIT. (2003). *The Future of nuclear power. An interdisplicinary MIT study*. http://web.mit.edu/nuclearpower/.
Price, R., & Blaise, J. R. (2003). *Nuclear fuel resources: Enough to last?* http://www.nea.fr/html/pub//newsletter/2002/20-2-Nuclear_fuel_resources.pdf.
Santos Ruesga, M. (2008*). Análisis económico de un proyecto de ampliación de la producción eléctrica nuclear en España*. http://www.foro-nuclear.org/pdf/Analisis_economico_proyecto_construccion_nuevas_centrales_nucleares.pdf.
Seko, N. (2003). *Aquaculture of uranium in seawater by a fabric-adsorbent submerged system*. http://www.ans.org/pubs/journals/nt/va-144-2-274-278.

Chapter 12
Helios and Aeolus

The king of winds, Aeolus, is invited for a month to the Aeolian Islands, and he gives Odysseus a present: a bag with all the winds inside, except the wind that can take him back to Ithaca. While he's sleeping, his men go through the bag thinking of the treasures it might contain, and they release all the winds. They reach the island of the Lestrigons, man-eating giants who kill and devour the crew of eleven ships.

The Odyssey, Book X, summarized

August in New York

In summer it's always hot in the Big Apple, but the summer of 2003 had been especially sweltering. Temperatures had been around 35°C during all the month, with humidity approaching one hundred percent. The higher the temperature, the lower the New Yorkers' thermostats, up to the point where you needed a sweater at the office, while outside delivery men, mail carriers and peddlers had to suffer the heat, together with the tourists brave enough to venture through the smoldering avenues.

The fourteenth of August had begun just like any other day. Sweat. I spent most of the day locked into my office, in the physics building at the campus of Columbia University, right in the centre of Manhattan, working on a science article whose publication had been delayed longer than it should have, till my computer screen suddenly went black. I didn't start to worry until an hour or an hour and a half later, when I began to hear voices and door slams in the neighboring offices. I stepped into the corridor. "It seems that electricity is out all over the city", one of the department secretaries told me, visibly disturbed. There were a lot of nervous people, huddling in groups and trying to phone. I noticed one of the managers of the building was prying into a locker, lighting into it with a torch. I asked him if I could help in some way. "The electric generator doesn't work", he mumbled. "I have to call the firemen, there are people caught in the elevator". Dave Schmitz, one of my colleagues, rushed by, heading for the fire escape. "It might be a terrorist attack", he cried at me, without stopping, as I wondered what he was doing. "I'm going home before this turns into a mousetrap".

He knew what he was talking about. Like several hundred thousands of New Yorkers, Dave lives in Brooklyn and works in Manhattan. In normal circumstances, the way back home used to take him about an hour. With all the traffic lights out and the crowds trying to get home before nightfall, it took him almost eight, and he could count himself lucky to have made it. Many others had to stay over with friends, or spend the night in hotels, in their offices, on the benches in Central Station, or just stay awake.

J. J. Gómez Cadenas, *The Nuclear Environmentalist,*
DOI: 10.1007/978-88-470-2478-6_12, © Juan José Gómez Cadenas 2012

In hindsight, it would have been worse if the power cut had happened at dusk, but the fact that there were still several hours of daylight left prevented panic attacks. When night started to fall, we knew it wasn't a terrorist attack, and the fire brigades had rescued almost all the people trapped in the elevators. The subway was another story. Juan Botas, my host in New York, was among the tens of thousands of travelers caught between two stations, in the dark, in a carriage crammed with people, which in 5 min turned into a 110°F (45°C) sauna. "After 2 hours the people started to lose their nerves," he told me later, "if it had taken them much longer to free us, there might have been a catastrophe."

And yet, catastrophes were few, considering the magnitude of the blackout, the longest in recent USA history—more than 29 hours—and also the most widespread, as it affected around 40 million people along the Northeast coast of the country. The sun went down on Manhattan with no lights to brighten it but the glowing snake of automobile headlights trying to escape from the island. George Pataki, the Governor, had already declared a state of emergency, and the streets were chock-full of police, though they weren't as necessary as we had thought at the beginning. There weren't any noteworthy outbreaks of street violence, nor serious riots. Quite on the contrary, a lot of people decided to lend a hand, among them plenty of citizens who stood in as traffic guards to mitigate the phenomenal traffic jams. When the going gets tough, the tough get going.

Juan and me met in the restaurant we used to frequent and joined the flow of accidental night owls. We got a free dinner; our place was one of those which decided to give away the food rather than let it rot in the freezer. Later we walked the streets lit by gas stoves, between buildings feebly outlined by candles and the occasional electric generator. We ended up in Central Park, gazing at a sight that most Manhattanites have never experienced: the calm glow of the stars above a dark city.

"It's beautiful", I sighed.

"For one night", my friend answered.

A System that Cannot Withstand any Faults

We have seen that the electrical network is an extremely complex system, where tens of high power stations and thousands of wind generators work together producing electricity *right at the moment* when the consumers (millions of homes, industries and services) require it.

Figure 12.1 shows the crudest simplified model of an electrical network we can picture. All the power plants together are condensed in a battery with a potential difference V across the terminals. All the users of the system have been reduced to one light bulb, connected to the battery, whose incandescent filament glows as it puts up resistance R to the flow of electrical current I. The relationship between the current flowing through the circuit, the resistance the circuit puts up and the potential difference are given by the famous Ohm's law:

Fig. 12.1 The simplest
model of an electrical
network

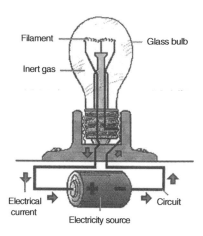

$$V = I \times R$$

which can be read like this: for a given potential difference, the smaller the
resistance of the circuit, the greater the current flowing through it.

In the electrical network, like in the smallest imaginable circuit, the problem is
keeping tension V at a constant value. In Fig. 12.1, as the battery runs down, the
light bulb will start blinking (V decreases, R remains constant and therefore I
decreases: less intensity, less brightness) until it finally goes out when V has
become so small it can't produce enough current.

The equivalent of the light bulb blinking in our simplified electrical network is
one of those power dips we experience from time to time, which usually last for
just a moment. But if there isn't enough tension for the current to flow through the
bulb (through the millions of light bulbs and electrical devices connected to the
network, each of them contributing to the total resistance R), a power cut occurs. If
the power cut lasts for a few minutes, we don't take it too seriously, except if we're
caught in the elevator or in the subway. If it lasts for several hours, we start to
worry about the frozen prawns. If it lasts for 6 hours, chaos is appreciable and the
press echoes the event, people call for scalps and write irate articles. If it lasts for a
day, it's likely that a state of emergency is declared.

Therefore, in order to rule out power cuts, the electrical network has to *be up to
demand* 24 h a day, 365 days a year. In the chilliest winter days and in the middle
of a sizzling summer, more power is consumed (heating and air conditioning) than
in autumn or spring, when temperatures are moderate. In Seville it's hotter than in
Santiago. By nightfall everybody turns on the heating at the same time. At seven in
the morning, all the water heaters all over the country are busy with the showers.
This means that the R in our circuit is constantly changing, and the total V of the
network has to adapt continuously. This is easy to understand, but in practice it's a
titanic task, with a lot of experts and sophisticated technology devoted to it.

Hydroelectric Energy

A thermal power plant is a reliable battery, which besides is able to modulate its tension easily as the charge of the network increases or decreases, given that electricity is generated by means of a giant turbine connected to an alternator that can spin faster or slower (the valves of the boiler are regulated to increase the flow of pressurized steam).

But hydroelectric energy is even more reliable. The power comes from the fall of a water mass in a riverbed, or in a reservoir held back by a dam, which turns the ubiquitous turbine. In fact, hydroelectric energy is in many ways the perfect power source. It's very efficient (90%, as compared to 40% for a thermal power plant) and can be easily regulated by opening or closing the floodgates, so it can be used to adapt to the changes in demand, to stand in for conventional thermal power plants or nuclear plants in case of temporary unavailability, and, within certain limits, as systems for storing excess power (wind power for example), pumping water to upland reservoirs so it can be used for peak demand.

Pumped-Storage Hydroelectricity

A pumped-storage hydroelectric power station has two reservoirs. The water stored in the lower reservoir can be pumped to a higher elevation during the times when excess power is generated, so it can later be reused to produce electricity.

In Spain one of the most impressive is the one in Aldeadávila (Fig. 12.2[1]), in the middle course of the river Duero, just at the border between Spain and Portugal, 7 km from Aldeadávila de la Ribera, in the province of Salamanca. Its height is 140 m and its surface 368 hectares. Water is pumped there from other reservoirs, like the ones in Almendra and Saucelle, so the reservoir is always full and it constitutes one of the "safeguards" of the national electrical system, being able to respond to peak demand, or to stand in for thermal power stations temporarily out of service.

In order to visualize how much energy this dam can store, let's assume that during the night (when demand is at its minimum) the pumps raise the water level in the reservoir by 3 m. The next day, during the hours of maximum demand, this mass of water slumps on the turbines.

How much energy has been stored in the 3 m of water at a height of 140 m? As we've already seen, Ep = m × h × g. If a surface of 360 hectares (3,600,000 m^2) has been raised 3 m, the corresponding water volume is $3 \times 3,600,000 =$ 10,800,000 m^3. Each cubic meter weighs a ton, so m = 108 million tons. When it falls, the water is accelerated at a rate of 10 m per second, *each second,* that is, 10 m/ s^2. If we multiply mass, height and acceleration, we find that the energy stored in the

---

[1] http://es.wikipedia.org/wiki/Archivo.HDR_Presa_de_Aldeadávila.jpg.

Fig. 12.2 The Aldeadávila dam, in the Duero River, one of the outstanding hydroelectric engineering works in Europe. Courtesy of Wikipedia

dam is 150×10^{12} J or 150 TJ, which is about 4 million kWh, enough to cover the demand of almost *half a million*.

Hydroelectric Power in Spain

The construction and use of the first hydroelectric power stations goes back to the origin of the electric industry. In 1882, the world's first hydroelectric plant was started in Appleton (Wisconsin, USA). It had the capacity to power—believe it or not!—not less than 250 lamps! 15 years later, the first large hydroelectric station came into use: the famous station at Niagara Falls, with 54 m height and ten groups totaling 50,000 horsepower to provide electricity for the city of Buffalo.

In Spain, the first hydroelectric stations were built towards the end of the 19th century. In 1901, 40% of the power plants in Spain were hydro powered, supplying a total power of 37 MW. In 1940, the power amounted to only 1,350 MW (more or less the equivalent of a modern thermal plant). In 2005 this figure had been multiplied by fifteen, reaching around 20,000 MW.

Since the middle 1960s, electricity generation has shifted increasingly to thermoelectric fossil fuel plants and later to nuclear stations, so hydroelectric power has decreased as a percentage. Nevertheless, the building of hydroelectric plants didn't stop until well into the 1990s. Consequently, hydropower plays an important part in Spain nowadays (contributing 10% to the electric mix, a higher rate than France, Japan, Germany or the USA), though our resources are limited compared with other countries such as Argentina (40%), Sweden (50%), Canada (60%), Venezuela (70%), and Brazil, Iceland, Norway, Uruguay and Paraguay, all of which get more than 80% of their electricity from dams.

But even this energy source, wonderful in almost all aspects, poses problems. Among them, we have to consider the environmental effects of the reservoirs and the fact that the accumulation of decaying biomass releases substantial amounts of the dreaded CO_2.

The safety of installations also has to be taken into account. In Spain, the Ribadelago dam (Zamora) burst 50 years ago, causing 144 deaths (five times more than the Chernobyl casualties) and blotting out the town. More recently, in 1982, the burst of the Tous dam killed 30 people and produced huge material damages. As I mentioned in Chap. 10, the Chernobyl tragedy has to be compared to others at a similar scale in order to gauge its significance, going beyond anti nuclear propaganda.

Finally, in contrast with the other massive energy sources, the hydroelectric production of a given system is subject to great variability from year to year due to differences in rainfall. In a drought-stricken year, hydroelectric power is below normal. Consequently, alternative installations (usually combined cycle, CCGT, plants) have to be built and maintained so that we can ensure a minimum of power during droughts.

In any case, in Spain water resources have already been put to good use for electricity generation. Though the potential of this technology has not yet been fully exploited, at the moment it is not foreseen to build new hydroelectric stations of more than 50 MW, but the enlargement of some existing plants is being considered. According to the current legislation on electric systems, major hydroelectric power stations are considered competitive and thus, in contrast with other renewables, don't benefit from any subsidies on the market price of power generation. However, the 2005–2010 Plan for Renewable Energies supports the construction of small hydroelectric waterfalls. Still, hydropower generation seems to have stalled at around 4,000 GWh, 50% below the amount foreseen for 2010.

Helios and Aeolus

In Greek mythology, Helios is the sun god, and Aeolus the lord of the winds. Helios represents the worship of our star, the heavenly body people have identified as giver of life since times immemorial. As ancient as the worship of the god is the rite of daily death and resurrection. Helios sinks below the horizon at dusk, leaving the Earth immersed in darkness during the long night hours. His absence also brings cold and the fear of the unknown. For millennia, men have sighed with relief each time they saw his rebirth at dawn.

Aeolus, on the other hand, is something of a jailor, his task is to discipline the wayward winds. There's a memorable scene in the *Odyssey* where, the coast of Ithaca already in sight, Odysseus' men open the bag of the winds and unleash a storm which blows them away from home again.

Both myths can be translated into modern terms by using the word *intermittence*. The sun shines for a limited number of hours a day, is not as warm in winter as it is in summer (besides days being shorter). Winds are variable in strength, from dead calm to hurricanes. And, as we have seen, electricity can't be stored at large scale (though hydro pumping provides a kind of relief), so it's impossible to "save" the wind energy of a strong Westerly day, or the midday solar energy, to use them at night or on wind-still days.

Obviously, electricity generation needs to be capable of continuously adapting to demand, and therefore the network has to be powerful enough to respond not just to the average consumption along the year, but also to any peak, like the ones that are frequent on a gelid winter day or during midsummer heat. Hydroelectric and thermal plants are capable of following demand peaks thanks to continuous electricity generation based on Parsons turbines, while wind generators have a poor response, due to the intermittent nature of wind.

In default of a viable technique for large-scale storage (something which is not in sight at the moment), the only solution to this problem is to rely on excess wind power or on thermal power stations working at idle (which is highly inefficient) to make up for an eventual lack of response to demand peaks. If we want the reserve thermal stations *not to release CO_2*, the obvious conclusion is, they have to be nuclear. It's worth stating it again:

> The large-scale development of wind energy requires the use of a sufficient number of thermal power stations to compensate for intermittence. The only type of thermal stations that can ensure the cleanness of wind energy—because they don't emit CO_2 either—is nuclear. From an environmental point of view, the expansion of wind energy doesn't make sense if it isn't accompanied by the corresponding nuclear expansion.

It's paradoxical, and it's a sign of the dualistic mindset so common in our times, that Greenpeace, an organization that firmly believes in wind energy,[2] opposes nuclear energy with equal determination, in spite of the obvious advantages that their association provides.

Photovoltaic Parks

Every second, the sun transforms four million tons of liquid hydrogen. Only a small fraction of the immense amount of energy radiated into space reaches our planet. Solar radiation, on a horizontal surface, depends on latitude; it's stronger near the Equator (around 2,000 kWh per square meter and year), and also in sunny desert areas like the ones so plentiful in Spain. Here, the average radiation per square meter and year is about 1,500 kWh, compared with 1,000 kWh in Northern Europe.

---

[2] A belief the author shares with them, by the way, in contrast with many other eco extremists who don't think wind generators are acceptable either.

The total energy consumption in Spain is about *300,000 million kilowatt hour* (300 TWh). Each square kilometer receives 1,500 million kWh (there's a million square meter in a square kilometer), so 200 square kilometer would suffice to cover the country's needs completely.

Hence the problem is not a lack in the resource (our country has an area of more than half a million square kilometers) but how to make use of it.

Direct conversion of sunlight into electricity is performed by means of photovoltaic panels, made up of a great number of cells where the photovoltaic effect transforms solar energy into an electric current.

The French physicist Edmond Becquerel (father of Henry Becquerel, who co-discovered radioactivity together with the Curies) is usually credited with the discovery of this effect, though the more general phenomenon (the photoelectric effect) was discovered and described by Heinrich Hertz in 1887. The theoretical explanation goes back to Albert Einstein in 1905. Contrary to what many people think, Einstein never received the Nobel Prize for the theory of relativity. Actually, he was awarded it for describing how a quantum of light, a photon, is absorbed by an electron acquiring its energy. When these electrons gain so much energy that they can be ripped out of a semiconductor material, an electric current is generated.

Until very recently, most solar cells were made of extremely pure monocrystalline silicon (that is, a block of silicon with just one unbroken lattice). This kind of crystal has no defects or impurities and is manufactured in bars, mainly for use in the microelectronics industry. The bars are cut into fine wafers, a dopant is added and metallic conductors are placed on each surface. The wafers are placed as tiles over a glass surface. Each wafer has a metallic contact to carry power to the grid. The life span of these devices is about 25 years. In the last years, thanks to intense R&D activity in this sector, new technologies have been developed, such as polycrystalline silicon (less efficient for conversion, but at a lower cost) and thin film technology, based on amorphous silicon, CIS (copper indium diselenide, CuInSe2), cadmium telluride (CdTe), etc. These developments are still at experimental stage; they are already being mass-produced, but volumes are still low.

Solar panels are arranged in modules that can be installed on the roofs of housing or industrial premises, or in solar parks (Fig. 12.3 above). On average, a solar plant generates 1 kW peak power for every 30 m^2 terrain (of which only 10 m^2 are occupied by panels, the rest of the surface is used for shade areas and the installations needed to extract electricity from the panels and channel it into the network, transformers, etc.).

Photovoltaic panels are connected in series so that they can offer a DC voltage between 450 and 600 V, and then they are connected in parallel to increase power. Spain has seen important, heavily subsidized investments in this sector in the last years. As a matter of fact, the World's three largest photovoltaic parks are in our country: Hoya de los Vicentes in Jumilla, Fuente Álamo, near Murcia, and Beneixama in Alicante (Fig. 12.3, below), each of them covering a surface of half a square kilometer (about 70 football fields). On the whole, 175 MW photovoltaic solar power is installed, placing Spain on second position in Europe, with a figure that triples the next country's (Italy, 58 MW) but is still far behind Germany (where not less than 3,063 MW have been installed!) Source (Foro Nuclear 2008).

Fig. 12.3 (*above*)
Photovoltaic panels; (*below*)
Photovoltaic park in
Beneixama, among the
largest in the World

Are photovoltaic parks commercially competitive?
To answer this question, we need to know:

1. how efficient solar panels are
2. the price of an installed solar kilowatt (in the photovoltaic case, it is usual to
   speak of "peak kilowatt", meaning the maximum installed power), including
   the costs of connection to the electric network, land acquisition, etc.

Both questions are related. The efficiency of solar panels depends on technology, which in turn has a bearing on the price. By way of example, let's consider the study carried out in Palma de Mallorca, mentioned in Chap. 3 in (Boyle 2004), comparing different technologies and concluding that all of them are around 10%. We have to multiply this figure by the "geographical" efficiency (the surface of the park which is actually occupied by panels), another 30%. So by multiplying solar conversion efficiency and geographical efficiency we conclude that photovoltaic installations of this size make use of a humble 3% of the available sun energy.

If the solar converters were totally efficient, we would need, as we have seen, a surface of 200 km^2 to produce all the electrical energy consumed in Spain. With an efficiency of 3%, the surface goes up to 6,000 km^2.

It's true that this area is just a tiny fraction of the surface available, but if we have a look at the picture of the park in Beneixama, bearing in mind that 12,000 parks like this one would be needed, we will be persuaded of the colossal installation we are talking about.

What about the price? The next table shows the main features of Beneixama solar park.[3]

Features of the solar park in Beneixama (Alicante, Spain)	
Nominal power	20 MWp (200 × 100 kWp)
Global radiation	1,934 kWh/m^2
Surface	500,000 m^2
Solar modules	100,000 units
Module surface	10,000 m^2
Electricity generation	30 GWh
Homes lit	5,000
Lifetime	25 years
Price	150 million Euros

In order to calculate the price of an electrical installation in a way that allows comparison with other installations, we divide the investment needed for its construction into the annual energy it provides. In the case of Beneixama, the investment amounts to 150 million Euros, and the annual electricity generation is 30 million kilowatt hour (30 GWh), so the price of the energy generated is 150/30 = 5 Euros per kWh.

Beneixama is one of the largest solar parks in the World. Let's compare its power and price with a newly built nuclear station. I have based my figures on a recent study by M. Santos Ruesga (Ruesga 2008), whose estimate for the inversion necessary to build a 1,000 MW reactor is 3,000 million Euros.

A nuclear station works for 24 h a day and practically 365 days a year, with only rare interruptions for refueling or servicing. Ruesga's estimation is 8,000 operating hours (from a total of 365 × 24 = 8,760, that is 10% of the time devoted to refueling and servicing), with 95% efficiency, so in one year we get 0.95 × 8,000 = 7,600 electricity hours. We multiply by the power of the plant (1,000 MW = 1 GW) and conclude that the annual energy generated is 7,600 GWh.

According to this, the cost of a newly built nuclear station would be 3,000 million Euros divided by 7,600 million kWh, amounting to 0.4 Euros per kWh. The price of fuel adds less than 5% to construction costs.

If we divide the cost of a solar park (5 Euros per kWh) into the cost of a nuclear plant (0.4 Euros per kWh), it turns out that the former is 12.5 times more expensive than the latter, electricity generation amounting to the same value.

---

[3] http//www.city-solar-ag.com/index.php?id=197.

Another important consideration is the number of solar parks like the one in Beneixama that have to be built to compete with a 1 GW power station. We just have to divide the 7,600 GWh generated by the nuclear plant by the 30 GWh produced by the solar park and we get... 250 Beneixamas!

In order to replace all the nuclear-generated electricity in Spain (55,039 GWh in 2007) by photovoltaic energy, 1,835 Beneixama-sized parks would be needed, spread over more than 900 km^2. They would cost 275,250 million Euros, a sum we could use to pay for 92 nuclear plants yielding almost 700 TWh: more than double as much electricity as we consume in a year.

Who Benefits from Investment in Solar Parks?

Am I saying that solar energy isn't viable? Not at all. Again, I would like to compare the situation with the beginnings of electricity. In my opinion, solar panel technology is currently at the stage of Edison's Jumbos, but it's rapidly progressing towards Parson's turbine, with continuous innovations like amorphous silicon, thin CIS wafers, silicon spheres and third generation photovoltaic cells based on nanotechnology. I am convinced that all this R&D will lead to a "solar revolution".

How long till then? It's hard to know. A lot of challenges remain, and on many fronts. Improving cell efficiency, cost reduction, manufacturing at a scale of hundreds and then thousands of square kilometers, solving the problems of network connection... If I had to place a bet, I'd say that the technology will start to be competitive at large scale in twenty or thirty years. Since 1980, the price of solar panels has decreased by a factor of 4. It has to come down by another factor of 10 for the "solar revolution" we crave for to become real.

Until then, applications of photovoltaic technology are plentiful, from providing electricity to farms, shelters and other remote sites, to powering small electric engines and automatic installations. In Africa, photovoltaic panels mean access to electricity, however precarious, for many people living in small villages who would otherwise have to go without it.

Subsidizing Kenyan farmers so they can buy solar panels is an excellent idea. Compelling all newly constructed buildings to include them on their roofs is a purely political measure that benefits nobody except the industry involved. The same could be said about subsidizing the construction of large solar parks (in Spain, all renewables are included in a special scheme and receive important subsidies). These are very expensive installations taxpayers have to finance and whose real advantages are ridiculous. A 100 MW CCGT plant would produce double as much energy with an investment ten times smaller, releasing relatively little CO_2 into the atmosphere.

If it's about developing know-how, this can be done with less ambitious installations. If you want to support R&D, it's better to invest directly in research through universities, CSIC (Spanish Council for Science) or the industry.

In the last seven years, Spain has gone from virtually zero to almost 450 GWh of installed photovoltaic energy. The Renewable Energies Plan aims to reach around 600 GWh. By then, the final investment will have been around 3,000 million Euros: this is the price of a nuclear plant that would have generated 12 times as much energy, taking up less space, releasing less CO_2 (see Chap. 11), and having a longer life-span. The numbers speak for themselves.

Thermo Solar Plants

A thermo solar plant is, as its name implies, a conventional thermal power plant (where electricity is generated from steam moving the ubiquitous turbine). The difference with coal, natural gas or nuclear plants is the fact that solar radiation is used directly to heat the water. Different technologies are employed. One example is PS10 in Sanlúcar la Mayor, Seville, Spain (Fig. 12.4), in use since March 2007. PS10 makes use of mobile mirrors to catch the sunlight (heliostats) and concentrates it onto a tower, heating the fluid that goes to the turbine. This plant is made up of 624 heliostats and a 115 m high tower, takes up around 64 hectares, has a power of 11 MW and generates 24.3 GWh a year.

In a thermal station, you need a continuous cycle of high pressure and high temperature steam. The way to obtain this in PS10 is by combining a system storing steam (large tanks where steam is stored at high pressure, which can work for about one hour) with a natural gas boiler that starts working at night, or with low sunlight.

In other words, a thermo solar plant is always something of a conventional thermal plant (that is, apart from solar energy it needs to burn fossil fuel, in this case natural gas). In fact, with conventional technology[4] the most efficient way to make these plants work is as hybrids: a solar park during daytime, a thermal plant at night. It's not a very orthodox solution, but it's quite convenient. Considering costs, these plants are three to five times cheaper than photovoltaic parks, so they seem, especially in their hybrid version, a more sensible option for the next decades, especially in countries with strong sunlight, such as Spain.

Wind Energy

Just as in the case of solar energy, Spain is second in Europe, after Germany, in the development of wind energy (Fig. 12.5[5]); (Foro Nuclear 2008). This is a booming industry in my country, with companies like ACCIONA, which from 2006 to 2007 doubled the number of wind generators (from 284 to 582), a figure that in turn is almost double as much as the number in 2005 (149). This impressive progression is represented in Fig. 12.6.

---

[4] See for example (Boyle 2004), Chap. 4.

[5] http://es.wikipedia.org/wiki/Archivo:Turbiny_wiatrowe_ubt.jpeg.

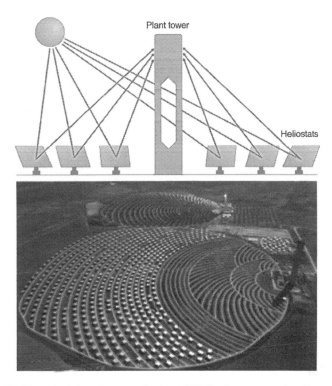

Fig. 12.4 Working principle and panoramic view of PS10 solar power station. *Source* (Abengoa Solar 2008)

Fig. 12.5 A wind park. Courtesy of Wikipedia

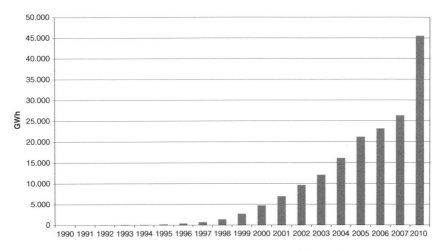

Fig. 12.6 The evolution of wind energy in Spain. *Source* (Foro Nuclear 2008)

Typically, the power of a turbo generator is between 1 and 3 MW. Its cost is highly dependent on the price of the turbines, which make up 70% of it and limit the life span to about 20 years. Then you have to add the costs of the construction works and the installation of electric lines to transmit the electricity.

Let's take the wind farm at Sierra Menera (Guadalajara, Spain) as an example. In 2006, the building costs were estimated at 45 million Euros,[6] plus 9 million to set up the transmission lines. The farm has 20 generators totaling 40 MW, so the cost would amount to 54/40 = 1.37 million Euros per MW. Similarly, the wind farm at Alcalá de la Vega y Algarra (Cuenca), recently inaugurated by Iberdrola Renovables, has a total power of 50 MW (25 turbo generators, 2 MW each) and needed an investment of 60 million Euros[7] (that's 1.2 million Euros per MW, in this case transmission infrastructure was already in place). If we average both figures, we get 1.3 million Euros per MW as the current price of wind energy in Spain.

Let us remind ourselves that we don't pay for the installed power, but for the electricity generated. In one year, the number of usable wind hours is around 2,000 (Ruesga 2008; Boyle 2003). Taking an efficiency factor of 95% for the turbines (Boyle 2004), it turns out that 1 MW wind power would yield an annual energy of $1 \times 2,000 \times 0.95 = 1,900$ MWh, or 1.9 GWh. The generation cost per GWh is 1.3/1.9, that is, approximately 0.7 million Euros: 0.7 per kWh, 75% more expensive than nuclear energy, and seven times cheaper than photovoltaic energy.

In view of these figures, it's not surprising that wind energy has undergone such an important commercial development, as it is the most competitive economically.

---

[6] http://www.labolsa.com/cronica/12574.

[7] http://www.eleconomista.es/economia/noticias/494828/04/08/Economia-Empresas-Iberdrola-Renovables-inaugura-un-parque-eolico-en-Cuenca-tras-invertir-60-millones.html.

In Spain it makes up 9% of the mix, so it's by far the most important renewable energy source.

Wind energy has lots of advantages. Installation is easy and investment costs per power unit makes it attractive—nuclear plants require a much higher initial investment and rigorous planning, while for a wind farm with a few turbo generators you just need to install the turbines and adapt or build the electric lines. Wind farms are easy to maintain and benefit from all the charm of renewables. They take advantage of a "fuel" that's available, whether we use it or not, and which is inexhaustible, plentiful and *free*. There's no safety issues, no CO_2 release, no waste whatsoever, and for the moment terrorists aren't paying attention to them. Compared with thermal stations, they're advantageous because they're clean and you don't have to import the wind. They're also superior to nuclear stations in terms of building complexity and safety. You don't need protection bunkers, or trained personnel to handle radioactive waste, etc. To sum up, if we lived in a world with wind blowing 24 h a day, 365 days a year, this would be the perfect way to generate energy, together with hydro power.

It's true that turbo generators aren't environmentally sound to a hundred per cent. Not everybody likes seeing the horizon full of blades spinning in the wind, and there are complaints about the risks they pose for birds, the noise they produce, interferences with electrical devices, etc.

In my opinion, these problems are negligible compared with the advantages. Wind generators, like electricity pylons, are part of the price we have to pay for civilization. We want electricity to be clean, cheap and convenient, and on top of it we demand that it be generated, or transmitted, by magic. Personally, I think wind generators are beautiful, above all when considering they're both an ancient idea (since the time of the Romans) and the first true success of industrial civilization in harnessing renewable energies.

But I'm afraid their rapid establishment—which has not been accompanied by the desirable educational campaign, a mistake that was also made in the case of nuclear energy—might turn against them. You hear more and more people complaining about the wind turbines, and it seems to me that the arguments they brandish hide the ugly NIMBY effect I dealt with in Chap. 9.

Why? Because, the same as in the case of waste deposits, wind turbines are perceived by the local population as something with no pay-off. They spoil the landscape (you need 2,000 turbines spread out over more than 200 square kilometers to replace a thermal stations, be it coal-fuelled or nuclear), they are noisy and, however beneficial they may be, we'd rather see them in the neighbors' backyard. The NIMBY attitude is shortsighted and selfish, but radical groups can take good advantage of it, and find that faint-hearted politicians pay heed to it. The legend of Aeolus comes in handy. No wind is more capricious and wayward than public opinion.

A 100% Renewable Mix?

To round up this analysis, I would like to comment on the proposal for a 100% renewable system made by Greenpeace, derived from a study by the University of Comillas *(Universidad Pontificia de Comillas, UPC)* (Greenpeace 2008).

The graph in Fig. 12.7 shows the electrical energy produced by each source, assuming the national consumption is 280 TWh. The first thing that strikes you is the fact that the sum of all energies is 500 TWh, almost twice as much as consumption. It's not a mistake. Lacking a reliable storing system for electricity, the only way to ensure sufficient energy when the resource is variable is to have plenty of it.

Let's see why. Suppose Spain has just two cities, Santiago and Seville, both with the same population number, and both of them get all their electrical power from wind turbines. Let's also suppose that most of the time it's equally windy in both cities, but from time to time there's wind only in one of them, and lucky enough, when it's dead calm in Seville, the wind howls in Santiago and vice versa.

Then we calculate like this: both in Seville and in Santiago we have to install double as much wind generators as we need, so that if there's no wind at all in one of the cities, the other can make up for it by generating twice as much power (and transmitting the electricity across the network built to this end). That is, we have to *double* the nominal power. If the wind was nice enough to always blow just in one city, we wouldn't be wasting any power (when the wind blows in Seville, Seville would have all the turbines running to meet both its own demand and the demand of Santiago, and the turbines would stop when the wind blows in Santiago). But most of the time, the wind blows in both cities, moves all the turbines installed and the power excess (which in practice can't be stored) is wasted. That's the most important reason why wind energy is still expensive. Building costs are proportional to the number of wind generators (that is, to installed power), while the electric energy they provide depends on variable winds (yielding an efficiency of around 20%).

And that isn't all there is to it. No sane person would risk basing all the system just on the wind, knowing that it may not blow in either of the cities. The electric network has to be based on a power source that is always available, and if there aren't enough pumping stations, you are forced to resort to thermal stations.

In the study carried out by the UPC, this thermal power is thermo solar, and it makes up almost 200 TWh!

Is it realistic to install 200 TWh of thermo solar energy? Taking PS10 as a reference, the only plant working in Spain in 2007, it doesn't look easy. PS10 covers half a square kilometer, it cost 35 million Euros and it generates around 24 GWh, that is, 0.24 TWh. In order to reach 200 TWh thermo solar power, you need 8,000 PS10, that's an area of 4,000 km^2, at a price of 280,000 million Euros, enough to build 93 nuclear stations yielding 706 TWh. Spread over 50 years, you'd have to build 160 PS10—totaling 125,000 heliostats—*annually*.

On top of it, this expensive and unrealistic proposal releases a considerable amount of CO_2 into the atmosphere. Let's not forget that a thermo solar plant

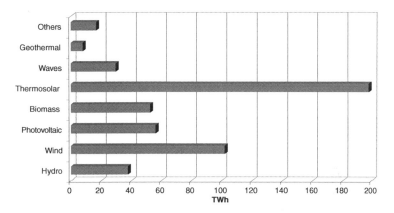

Fig. 12.7 Proposal of electrical mix based on 100% renewable energies. *Source* (Greenpeace 2008)

works on sunlight during the day, on fuel at night. Assuming that a thermo solar plant is working on natural gas 50% of the time, we would be consuming 100 TWh according to the proposal, or 50 TWh if it was working just 25% of the time.

Concerning photovoltaic energy, the study by UPC assumes 50 TWh, equivalent to 1,7000 Beneixamas at a price of 225,000 million Euros (85 nuclear stations, 646 TWh), covering around 800 km^2.

For wind energy, the UPC's assumption is 100 TWh, compared to the current figure of about 30. This is the only case where the number of wind generators is just multiplied by three, something much more feasible than the other two proposals, at a price of 70,000 million Euros, which on the other hand would be enough to build 47 nuclear stations, providing 357 TWh.

For the rest of energies under consideration, except hydropower, the pattern is similar. Getting 50 TWh from biomass implies multiplying the current production by seven, and getting 30 TWh from the waves (at the moment there's no system of this kind in operation) seems equally uncertain.

But in fact the situation is even worse, because in 50 years electricity consumption won't be at 300 TWh, but will have risen to 600 TWh (based on a moderate growth rate of 1.5% annually, half the current growth rate). The need to oversize leads to planning for 1,000 TWh, with costs and technical difficulties increasing correspondingly.

These quick calculations don't mean to prove that renewable energies aren't viable. I'm not against them—quite on the contrary, I'm a fervent supporter of developing renewables, without forgetting their particular features and limitations—but against a dogmatism that leads to such curious extremes as on the one hand praising the need for saving energy and on the other suggesting a mix that wastes half the electrical power due to the variability of the resource.

In my opinion, a mix based on wind energy, hydropower, hybrid thermo solar and nuclear is sensible, economically feasible and environmentally friendly. The tiresome repetitions of the litany notwithstanding, there's no antagonism between nuclear and renewables, quite on the contrary. Renewable energies *need* nuclear energy, the only alternative to fossil fuels that doesn't release CO_2 and can be generated continuously in large amounts, thus making up for the renewables' Achilles' heel, the unavoidable intermittence of the resource. It's a pity that certain environmental organizations, whose moral responsibility should be on par with their financial resources and their decisive social influence, seem unable to reconsider their dogmas.

References

Abengoa Solar (EMPRESA) (2008). *Tecnologías de torre para centrales termosolares*. http://www.abengoasolar.es/sites/solar/es/tecnologias/termosolar/tecnologia_de_torre/index.html.
Boyle, G. (2003). *Energy systems and sustainability*. Oxford: Oxford University Press.
Boyle, G. (2004). *Renewable energy*. Oxford: Oxford University Press.
Foro Nuclear (2008). http://www.foronuclear.org.
Greenpeace (2008). *Informes renovables 100%*. http://www.greenpeace.org/espana/reports/informes-renovables-100.
Santos Ruesga, M. (2008). *Análisis económico de un proyecto de ampliación de la producción eléctrica nuclear en España*. http://www.foronuclear.org/pdf/Analisis_economico_proyecto_construccion_nuevas_centrales_nucleares.pdf.

Chapter 13
At the Crossroads

> *Two roads diverged in a yellow wood,*
> *And sorry I could not travel both*
> *And be one traveler, long I stood*
> *And looked down one as far as I could*
> *To where it bent in the undergrowth*
>
> Robert Frost, The Road Not Taken

Waiting for Godot

Every year, the International Energy Agency publishes a study about energy in the world, called WEO (world energy outlook). In the 2008 version (IAE 2008) three scenarios are considered. In the so-called reference scenario, the world goes on with business as usual. Electricity generation increases from almost 19,000 TWh (trillions of kilowatt-hour) in 2007 to more than 33,000 TWh in 2030. China overtakes the USA in electricity consumption and equals Europe, while India overtakes Russia (Fig. 13.1).

In the reference scenario, electricity consumption is still based on fossil fuels, and in fact the ratio of coal in the production of electrical power goes from 41 to 44%, leading to an increase in CO_2 emissions, which go from around 30,000 million tons in 2007 to 41,000 million tons in 2030 (Fig. 13.2).

In Europe and the USA emissions increase very slowly and finally stabilize, but China shatters all hopes of moderation. In India, CO_2 release speeds up likewise.

And yet, measured in tons per capita, the USA still emit considerably more than China, though the Asian giant will have caught up with Europe in 2030. The global CO_2 increases at a rate that suggests an excess in carbon dioxide concentration of 700 ppm towards the end of the century, a rise in temperature of about 6 degrees… and a guaranteed catastrophe for our civilization.

The IEA's reference scenario is hair rising in several aspects. It reveals a world incapable of anything but steering towards collective, premeditated suicide, and proves pessimists like Lovelock (2007) right.

> If we fail to concentrate our minds on the real danger, which is global heating, we may die […] Our goal should be the cessation of fossil-fuel consumption as quickly as possible […] To undo the harm we have already done requires a program whose scale dwarfs the space and military programs, in cost and size. We live at a time when emotions and feelings count more than truth, and there is a vast ignorance of science. We have allowed fiction writers and green lobbies to exploit the fear of nuclear energy and of almost any new science, in the same way that the churches exploited the fear of Hellfire not so long ago. […] We cannot afford to wait for Godot.

J. J. Gómez Cadenas, *The Nuclear Environmentalist,*
DOI: 10.1007/978-88-470-2478-6_13, © Juan José Gómez Cadenas 2012

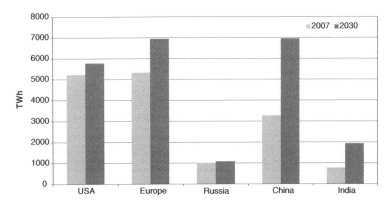

Fig. 13.1 Increase in electricity consumption in several countries. *Source* (IEA 2008)

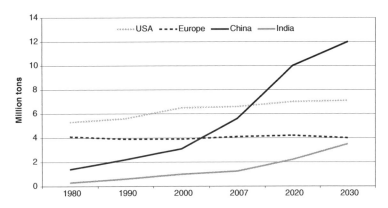

Fig. 13.2 CO_2 emissions in several countries. *Source* (IEA 2008)

The Challenge of the Century

There's No Magic Solutions

Are there any solutions? On paper, there are. I have already pointed out that a combination of nuclear energy and renewables (mainly wind energy and hydro-power, progressively including solar, geothermal, tides etc., as they're being developed and their costs go down) could provide an alternative scheme without release of CO_2. In a more distant future (at least 50 or 60 years, perhaps a century) we could use nuclear fusion energy together with hydrogen for storing energy and as a fuel for transport... These are exciting possibilities that, together with other futuristic alternatives, would need a book for themselves. The topic won't go out of fashion. In the year 2030, there may be more buses, and some cars, running on liquid hydrogen or fuel cells, but transport will keep depending on gasoline and

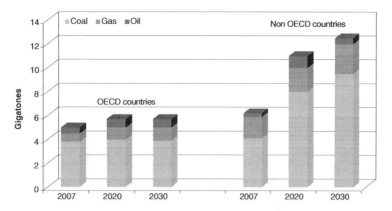

Fig. 13.3 Contribution of the various fossil fuels to CO_2 emissions. *Source* (IEA 2008)

diesel. If the price of fuel goes up and so do the taxes on CO_2 emissions, hybrid electric vehicles will have lower emissions and we might see more electric cars and, hopefully, more bicycles in the cities. Regarding nuclear fusion, a new Manhattan project with climate change as the enemy is called for; otherwise, I doubt that my generation will live to see this miracle.

The Magnitude of the Problem

Figure 13.3 gives us an idea of the scale of the problem we are dealing with. According to the reference scenario, in the year 2030 the power stations of the OECD countries will be releasing almost 6 gigatons of CO_2, and the countries outside the OECD, led by China, more than double that much, altogether totaling 18 gigatons of carbon dioxide, of which 14 (78%) correspond to coal consumption.

This is only half of the problem. The other (large) half (23 gigatons) is mainly due to transport.

How can we get rid of these tremendous emissions, or at least reduce them significantly? We're talking about a colossal figure, 41 billion tons of carbon dioxide. I have no answer as regards transport, and I don't know if anybody has. One possibility would be high petrol prices, combined with sound policies that impose an increasing tax on exhaust fumes, persuade the citizens to replace their SUV by a Smart, a new generation hybrid or an electric car. On the other hand, this scheme can't work without a massive investment in public transport infrastructures. Lovelock isn't exaggerating when he calls this task titanic. The automobile industry has been the most important in the twentieth century, and it was often developed to the disadvantage of public transport. Educating citizens isn't enough. You just have to catch the Madrid underground at peak hour to understand that fundamental reforms are needed to allow a greater flow of travelers.

Let us examine the implications of doing away with the CO_2 emissions due to coal in the process of electricity generation. If we achieved that, total emissions would be lowered by 30%, which is certainly not enough, but represents a preliminary victory.

According to the reference scenario, by 2030 we will be consuming 15,000 TWh generated from coal. It should be our goal to replace this consumption by renewables in 20 years.

Each 1,000 MW nuclear plant provides 7,600 GWh (7.6 TWh) per year. In order to replace all coal-fired thermal power stations between 2010 and 2030, it would be necessary to build 2,000 nuclear plants ($2,000 \times 7.6 = 15,200$ TWh) in 20 years, that's 100 plants per year in all the world.

Is this out of the question? France built 56 nuclear plants in 15 years, at a rate of almost four a year. This country generated 566 TWh in 2007 (compared to globally 19,894 TWh). Assuming that Europe (here we're including Russia and the ex USSR countries), the USA, China, India, Japan and South Korea could keep building plants at the same rate as their consumption increases it would turn out that four French nuclear plants a year is equivalent to 2 annually in Spain, 36 in North America (including Mexico and Canada), 36 in all of Europe, 24 in China, 4 in India, 8 in Japan and 2 in South Korea. Altogether that's 112 power stations, without counting the countries in the Persian Gulf, Africa or South America.

I don't mean to say this distribution is realistic, I just want to convey an idea of what a challenge it is to fight climate change.

Anyway, if France has managed to build 4 plants a year for 15 years, why shouldn't Spain build half as many; and China, one of the countries that have already launched an ambitious nuclear program, could as well build twenty of them. The European share doesn't seem out of the question, given the great number of powerful economies in the region (Germany, the UK, Italy and Russia are certainly able to build 4 or 5 nuclear plants yearly each). Many other countries I have not included (South Africa, Brazil, Argentina, Chile) will also contribute to some degree.

Finally nuclear energy isn't the only weapon to fight climate change. If wind energy (and hydropower wherever possible) was developed with equal enthusiasm, it could attain a share of between 2,000 and 5,000 TWh in 20 years. And then, if power stations fuelled with natural gas were progressively substituted by hybrid thermo-electric plants, we would make better use of the resource (sun during the day, gas by night) and emissions would be reduced further.

The global investment needed to carry out this ambitious energy Marshall Plan might not exceed a billion Euros annually, that is, an amount equivalent to the Spanish GDP. It doesn't seem excessive compared with the mammoth amounts that have been squandered on purely speculative ventures in these years of wild capitalism, and in fact this investment might be just what is needed by a world on the brink of the great depression. Great investment capable of generating jobs, added value, energy... and future.

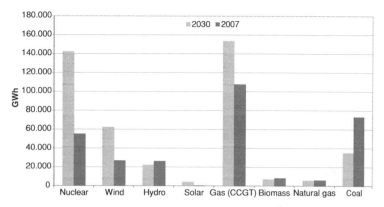

Fig. 13.4 Spanish mix for 2030 compared with the mix in 2007, according to Ruesgas's plan. Nuclear, wind and gas goes up, coal goes down. *Source* (Ruesga 2008)

What About Spain?

You can rightly argue that the figures I've given are not based on any trustworthy study. So let us compare this with the plan by economy professor Ruesga (2008), a thorough and well-documented study which advocates the construction of 11 new nuclear plants in Spain in the next 20 years, showing that this enterprise isn't just feasible, but beneficial for society.

If it were carried out, the electric mix would be made up of 33% nuclear energy, plus 23% renewables (with not less than 14% wind power), 36% natural gas and only 8% coal. Figure 13.4 shows how global electrical power increases (by almost 50%) while the absolute contribution of coal goes down. To achieve this, the nuclear contribution, as well as wind and gas, have to rise drastically. *Combining them all* is what allows lowering CO_2 emissions (let's not forget that natural gas releases half as much as coal).

This plan costs 33 billion Euros, which doesn't seem an exorbitant amount in view of the astounding figures that went up into smoke in the 2008 crisis, and isn't preposterous either in the context of government investments to fight the crisis. You could say that this is the right moment to contribute with great public works, stimulate industry, create jobs (more than 200,000 a year for 20 years, according to Ruesga's exacting analysis), recover lost know-how and prevent the remaining from being lost, reduce our deplorable energy dependency and, above all, release less CO_2.

The current policies supporting renewable energies appear correct to me, except for the excessive funding of photovoltaic installations (a situation that has already begun to change) when the money should go to R&D. In my opinion, promoting nuclear energy in Spain is a challenge for the future, as long as the industry is up to it. It would have to be accompanied by strengthening R&D in the energy sector.

Fission (fifth generation), fusion and renewables. By the way, the energy sector is among the ones most in need of deep reforms if we want to cease being a country with nothing to offer but sun and property building.

The Lessons of the Past

Developing the great energy systems is a long, cumulative process, a fact that sometimes goes unnoticed in the public discussions about the pros and cons of the various alternative energies. I have chosen a time span of 20 years as an example to convey an idea of the challenges posed by a nuclear program to fight climate change and do away with the dependency on fossil fuels, but in practice this program will have to be developed over a period of at least half a century, with successive generations of continuously enhanced nuclear reactors that will replace the previous ones (as has happened and will continue happening in the renewable energy sector, as proven by the spectacular improvement of the wind turbines in the last decade).

The antagonism between nuclear and renewables is an absurdity stemming from the dualistic mindset that has taken hold of our society and especially of the rows of eco activists. It's wrong to pit alternative energies against each other; it would be more sensible to consider them as allies in the great war against an ever more powerful Dark Lord who might well extinguish, in one or two centuries, our dear and wretched global village.

A Rose

Perhaps in the future our century will be remembered for its war against climate change, and I hope it will be the remembrance of a victory. To win the war, more R&D is essential, both in Spain and in the rest of the world. In the long run, the nuclear option won't succeed unless we learn to recycle the waste correctly and to transmute the trans-uranians. If we strive for an expansion like the one I've pictured, reactors have to be mass-produced, their designs standardized, costs reduced and safety improved. It's indispensable to regain the trust of public opinion.

Trees have to grow on the chimneys of the nuclear power stations, and flowers under the blades of the wind turbines, quite literally. R&D will sooner or later lead to efficient systems for harnessing solar energy and with time, to developing nuclear fusion.

In the last decades there has been a lot of talk about "reducing man's ecological footprint" on the planet, a concept introduced by the influential book *The Limits of Growth* (Meadows et al. 1972). I agree with the idea set forth by the authors of this important work, which is idolized by some, scorned by others and, as it seems to me, misunderstood by most. It's true that many of the predictions of Dana

Meadows and her co-authors didn't come true, especially those related to the impending scarcity of raw materials, but the moral and philosophical lessons of the work more than outweigh the minor numerical flaws. The later versions of the book deliver a masterful account of the profound contradictions of a world obsessed with growth and at the same time deeply unjust. The central concept the work hinges on is collapse, the likely fate of a society exceeding its limits. Some of the limits considered by the authors (for instance, the availability of raw materials) have turned out to be much farther away in time than they were thought to be in the 70 s, but on the other hand we may well have already trespassed the limit related to emissions of greenhouse gases.

The solution proposed in *The Limits of Growth* consists in reducing our ecological footprint, the human impact on the planet. These are more or less Lovelock's ideas. It's difficult not to agree, but it would be hypocritical to aim at a lighter footprint by keeping three quarters of mankind in poverty.

In my view, the only way to respect nature is to develop science and technology so that they allow us to stop exploiting it. There's no way to feed 9 billion people with "natural" farming methods (that is, without the use of herbicides or fertilizers), but in a not too distant future we may be capable of handling such enormous amounts of energy that we can *synthesize* foodstuffs and return the fields to Gaia. Similarly, I don't think the solution to the energy crisis is saving, given that only 20% of the world's population is in a position to save, while 80% need more energy to escape from destitution. I believe we have to learn how to generate all the energy we need without making the planet pay a price for it. Biomass may be a renewable energy, but I prefer to obtain hydrogen from nuclear energy rather than cut down trees, or have hectares and hectares of farmland sown with corn to "harvest" biofuel, one of the most inefficient and least ecological methods I can think of to obtain energy, in spite of its green credentials.

There's nothing natural, desirable, or morally superior about poverty, nor about overabundance and squandering. So science and technology are not enough. More urgently than ever, we need a moral, ethical, ecological code.

Perhaps we need a miracle.

Miracles are sometimes within arm's reach if we know where to look.

I would like to conclude by remembering the young man who had traveled across half of Europe to become a disciple of the great Paracelsus and was unable to see beyond the empty room without distillers and retorts, and beyond the ancient man's frail bones and tired look.

This short sightedness was the reason why he returned to his dull existence and did not witness what happened in the alchemist's poor hut.

When the wannabe disciple leaves after throwing the rose into the fire, the wizard sits down on a humble chair and gazes into the embers, for a long time, immersed in thought.

Finally, he utters a word. And the rose rises from the ashes.

References

IAE (International Energy Agency). (2008). World energy outlook. http://www.worldenergy
    outlook.org/2008.asp.
Lovelock, J. (2007). *The revenge of Gaia: Earth's climate crisis and the fate of humanity.* UK:
    Penguin Books. ISBN: 0141025972.
Meadows, D. H., Meadows, D. L., Randers, J., & Behrens III, W. W. (1972). *The limits to growth.*
    Universe Books. (The book has been updated in two subsequent editions, 1992 and 2002).
Santos Ruesga, M. (2008). *Análisis económico de un proyecto de ampliación de la producción
    eléctrica nuclear en España.* http://www.foro-nuclear.org/pdf/Analisis_economico_proyecto_
    construccion_nue-vas_centrales_nucleares.pdf.

Chapter 14
Fukushima, or the Black Swan of Nuclear Energy

By its own merits, the great earthquake that hit Japan on 11 March 2011 would have qualified as one of the worse disasters of recent times. With a magnitude of 9.0 MW, it was the most powerful known earthquake ever to have hit Japan, and one of the most powerful in history. It released a surface energy 2×10^{17} Joule. Enough, if harnessed, to power a city the size of Los Angeles for an entire year.

The earthquake triggered a powerful tsunami which brought destruction along the Pacific coastline of Japan's northern islands, killed some 20,000 people and devastated entire towns, with waves reaching almost 40 m in some coastal locations.

The damage caused was enormous. Some towns were reduced, literally, to rubble. Before-and-after satellite photographs of devastated regions show immense damage to many regions, with a grand count of more than 45,000 buildings destroyed and some 150,000 damaged, plus about a quarter of million cars and tracks destroyed. Cost estimates range well into the tens of billions of US dollars.

Many energy-production systems were also damaged, including irrigation dams, electric transmission lines, oil refineries and a liquefied natural gas plant.

But the most famous consequence of the earthquake and the subsequent tsunami was the worst nuclear accident of the last 25 years. Indeed, the name Fukushima has joined that of Chernobyl to name the dangers of nuclear energy.

The Fukushima Daiichi plant comprised six separate boiling water reactors designed by General Electric (GE), and maintained by the Tokyo Electric Power Company (TEPCO). At the time of the quake, reactor 4 had no fuel and reactors 5 and 6 had been shutdown for planned maintenance. Reactors 1, 2 and 3 shut down automatically after the earthquake. Thus, unlike Chernobyl, Fukushima was not a criticality accident. Rather it was a full scale demonstration of the main weakness of the LWR, namely: *a light water reactor needs external power for cooling after a shutdown.*

The reason for this is that radioactive decay continues after the chain reaction has been stopped. During the first weeks after a scram (shutdown) the energy deposited by the alpha particles, electrons and high energy gammas releases in those decays, results in an enormous source of power, which is dissipated in the

J. J. Gómez Cadenas, *The Nuclear Environmentalist*,
DOI: 10.1007/978-88-470-2478-6_14, © Juan José Gómez Cadenas 2012

form of heat. In order to evacuate it, a continuous flow of cool water must be forced inside the reactor vessel and circulated through the reactor core. Pumping water, in turn, requires electric power. When the earthquake hit the Fukushima plant, emergency generators came online to power the systems needed for cooling, but the tsunami broke the reactors' connection to the power grid, initiating an accident that resulted in the meltdown of the reactors' cores.

The nuclear accident immediately became a media phenomenon. It occupied the headlines of all major newspapers and prime-audience news channels for entire weeks, essentially eclipsing all the other disasters caused by the tsunami. In some sense the media attention was well justified, given the magnitude of the catastrophe, which included violent hydrogen explosions and the feared meltdown of the reactors' cores. Even more fearsome for the public were the large radioactive releases that led Japan's government to impose a 20 km radius evacuation around the plant. On the other hand, many observers felt that the coverage of the accident, in particular from foreign media was "excessive" and too sensationalistic, resulting in widespread panic and taking away public attention from the larger-scale disaster. One commentator wrote:

> People in California start searching for iodide pills on the internet and there are already people voicing worries about whether Japanese cars are now all going to be radioactive. But worst of all, the inordinate and sensationalist attention given to the reactors by American and other media has taken attention away from where it should be: on the likely nearly 20,000 people who died in the quake and tsunamis, on the nearly 400,000 homeless people, and on the immense suffering this has caused for Japan as a whole.

Conversely, some of the Western media have criticized Japanese newspaper, official sources and TEPCO for offering too little or too biased information. For many, this accident has marked the end of an announced "nuclear renaissance". Certainly, some countries, like Germany, Italy and Switzerland have reacted to the accident with a clear move away from nuclear power, whose future is also quite uncertain in Japan.

The Fukushima Daiichi Nuclear Power Plant

The Fukushima I—also known as Fukushima Daiichi or "number one"—NPP was a very powerful NPP, delivering a combined power of 4.7 GW, thanks to its 6 BWR (boiling water reactors) units. It was also rather middle-aged. The first unit had been commissioned in 1971 and the last in 1979. Thus all the reactors had been operating for more than 30 years (unit 1 for 40 years). Five of the 6 reactors (units 1–5) had Mark I containment type. Unit 6 was equipped with the more modern Mark II containment type.

All LWRs use water as a coolant and as a moderator. This has a good and a bad side:

The good side is that the probability of a criticality accident is very small, due to a powerful negative feedback loop. Recall how it works: if the chain reaction goes out of control the temperature of the fuel (and thus of the water) increases, eventually causing water to boil. At this point the neutrons do not find any more hydrogen atoms to collide with. Without collisions neutrons do not slow down, and without slow neutrons the chain reaction stops. Therefore there is no criticality accident.

The bad side is that water needs to be continuously injected inside the reactor, well after the chain reaction has stopped, in order to keep the reactor core cool. The core is very hot due to radioactive decay, and, without a fresh supply the water filling the pressure vessel will boil, increasing the pressure inside the reactor and exposing the fuel elements, which in turn become even hotter. Once the temperature rises above some 1,800°C the zirconium of the fuel rods will melt, and above 2,700°C, it will be the turn of the uranium oxide pellets. Eventually the whole fuel structure will collapse, dropping into the bottom of the reactor, where the remaining water may even produce local a limited criticality (as the melted fuel finds water moderation may re-start a limited version of the chain reaction). This is the picture of the feared core meltdown.

The bottom line is that the LWR is protected by *physics* from a criticality accident, but human engineering is needed to prevent what is generically known as LOCA (Loss of Cooling Accident). Physics laws are immutable—water will always boil at a given pressure and temperature—but human engineering may fail. To pump water into a LWR reactor core one needs power. In the event of a generalized power failure, autonomous power systems, such as Diesels generators and DC (direct current) batteries are needed. In the Fukushima Daiichi NPP, those diesel generators and DC batteries were located in the basements of the reactor turbine buildings. The location strictly followed General Electric specification design, but ignored what appears, a posteriori, almost as an obvious fact. In a tsunami area, a higher level location would have been a must.

Radioactive decay is a source of power. This power decreases quickly with time, as the most radioactive elements decay, but it can be as high as 7% of the reactor power, right after the *scram* (reactor shutdown). This means about 70 MW for a typical, 1,000 MW machine. The heat due to radioactivity decreases to about 0.5%, or some 5 MW 10 days after shutdown.

Recall that a typical home electrical heater has a power of about 3 kW. Thus, the heat produced by the radioactive decay right after the scram is roughly that of 20,000 such devices. Even 10 days after shutdown, the radioactive power inside the reactor equals some 2,000 home electric heaters.

To evacuate this heat one needs to pump cool water into the reactor core. The water circulates among the fuel elements and heat is transferred to the liquid. The heated water exits the reactor and circulates through a cooling loop before entering the core again.

To pump the water into the core, power is needed. If no power is available, heat cannot be evacuated. This is exactly what happened in 3 of the reactors at the Fukushima I NPP.

The Accident Step by Step

On March, 11, 2011, at 14:46 the earthquake hits Japan. All the reactors at Fukushima scram automatically, suffering little damage. The electric grid shuts down, but the autonomous diesel engines start operation, guaranteeing the needed water flow. At this point the reactors are stable.

At 15:41 the tsunami hits the plant. The installation is prepared to withstand a wave of 6.5 m. The height of the tsunami exceeds 7 m, flooding the diesel engines. As a consequence, power is lost in all the plant except for the DC batteries. The batteries are emptied after a few hours. The water in the reactor starts to heat and the pressure from vapor increases. Pressure is alleviated by discharging water vapor into the wetwell. As a consequence the level of water in the core starts to descend.

Eventually the core is exposed. The fuel elements start to over-heat, but structural damage to the fuel is small while there is less than 50% of the core exposed. However, once 2/3 of the core becomes exposed, the fuel elements start to swell and break liberating fission products. When the temperature of the fuel elements exceeds 1200°C, the zirconium tubes that hold the fuel pellets start to burn in the water vapor atmosphere. The reaction is exothermic (liberates heats), and contributes to accelerating the heating of the core. Thus, a positive feedback loop is set in motion. Furthermore, the reaction produces a large quantity of hydrogen.

The temperature keeps raising: at 1800°C, the zirconium tubes start to melt; at 2400°C, the structure collapses to the bottom of the reactor and a pool of radio-active fused debris is formed; at 2700°C, the fuel pellets themselves melt.

At this stage, the fission products are still isolated from the environment by the concrete building, which is designed to withstand an internal pressure of 4–5 bars. However pressure inside the reactor has increased above 8 bar, by the combined effect of hydrogen production and water boiling. This forces the depressurization of the containment building, by liberating radioactive gases and aerosols to the environment. However, this is the only solution left to reduce pressure (and explosive energy) inside the containment building.

The hydrogen released is flammable, and hydrogen explosions follow the depressurization in units 1 and 3. In unit 2, there were also hydrogen explosions inside the reactor building, with disastrous consequences. Damage to the reactor vessel—which at this stage contained highly radioactive water—uncontrolled release of radioactive gases, and release of fission products.

The accident was eventually stopped, first by filling the reactors with seawater and later by using portable pumps. The radioactive doses in the site, particularly after the explosions in unit 2 were very high, reaching the 400 mSv/h. For comparison, the total dose that a person receives on average every year is about 3 mSv. These high levels of radiation made operations inside the plant very difficult and exposed the workers to severe doses.

Fukushima and Chernobyl

The Fukushima accident was very different from the accident at Chernobyl. In Japan, an earthquake followed by a tsunami provoked a massive LOCA that resulted in a massive core meltdown in four of the 6 units operating in the plant, but the integrity of the reactors containment building was largely preserved. At Chernobyl there was a criticality accident that destroyed the (almost non existent) containment and threw the reactor guts into the environment.

As a consequence, the release of aerosols (e.g., small radioactive particles) was much smaller than that at Chernobyl. A large fraction of the contamination was in the form of rare gases that diffuse quickly in the atmosphere. Overall, the radioactive release was about a factor 10 smaller in Fukushima than in Chernobyl.

However, like in Chernobyl, a large area (20–30 km around the plant) has been banned (it remains to be seen when the excluded area will be habitable again), populations have been displaced, and fear of radiation has been widespread both in Japan and elsewhere. Furthermore, the accident of the reactors was not the only problem in Fukushima. Used, radioactive fuel was stored in pools in the top of the reactor building, and the accident caused those fuel elements to be exposed to the air, causing further radioactive releases.

The LWR Achilles' Heel

A full analysis of the Fukushima accident including such delicate matters as the evaluation of how well it was managed or whether there was enough transparency in the information offered to the media and public is outside the scope of this book.

However, the accident reveals two crucial issues that need to be considered in any debate about nuclear energy.

The first and most important one is the recognition of the LWR Achilles's heel, the need for forced water pumping to avoid a core meltdown. The second is the practical demonstration that outdated technology can make a bad accident worse. This was the case with the fuel pools, which are protected inside the reactor building for the Mark II design but not for the older Mark I design.

One way to formulate the same issues is to consider any of the new LWR designs, such as the EPR, ABWR or APR and ask what would have happened if the Fukushima NPP had been equipped with those reactors rather than with its aging BWRs.

Going even further, one should ask whether it is reasonable to plan a large expansion of nuclear power based on LWRs, even if those are brand new EPRs or other safer and improved models. Perhaps the most important lesson from Fukushima is that any future, large nuclear expansion will need to be based on the newer, safer designs being developed by the Generation IV initiative.

19630612R00094

Made in the USA
Lexington, KY
29 December 2012